CONTENTS.

PREFACE.

THE following papers on the Sanitary Condition of Hospitals, and on Defects in the Construction of Hospital Wards, by Miss Nightingale, were read at the Liverpool meeting of the National Association for the Promotion of Social Science, in October, 1858, and are now printed in terms of a resolution of the Association.

The publisher has thought that it would further the objects of the Association if the evidence given by Miss Nightingale before the Royal Commission on the Sanitary State of the Army were reprinted, and he has added it accordingly. That evidence not only contains information of public importance, as regards the army and its hospitals during the late war with Russia, but gives the results of many years' experience with regard to the principles of hospital construction and organization.

In order to make the information on hospital construction as complete as possible, three papers from the *Builder*, of August 28th, and September 11th and 25th, 1858, have been also reprinted, by the kind permission of Mr. George Godwin, who has paid much attention to the subject. They present similar views on the architectural arrangement of hospitals.

SANITARY CONDITION OF HOSPITALS

AND

HOSPITAL CONSTRUCTION.

Two Papers read before the National Association for the Promotion of Social Science.

PART I.

FEELING very desirous of contributing whatever aid I can to improvement in Hospital construction and administration—especially at this time, when several new hospitals are being built—it has occurred to me to transmit a few notes on defects which have come under my own observation in an extended experience of these institutions.

No one, I think, who brings ordinary powers of observation to bear on the sick and maimed can fail to observe a remarkable difference in the aspect of cases, in their duration and in their termination in different hospitals. To the superficial observer there are two things only apparent—the disease and the remedial treatment, medical or surgical. It requires a considerable amount of experience, in hospitals of various constructions and varied administrations, to go beyond this, and to be able to perceive that conditions arising out of these have a very powerful effect indeed upon the ultimate issue of cases which pass through the wards.

It is sometimes asserted that there is no such striking difference in the mortality of different hospitals as one would be led to infer from their great apparent difference in sanitary condition. There is, undoubtedly, some difficulty in arriving at correct statistical comparison to exhibit this. For, in the first place, different hospitals receive very different proportions of the same class of diseases. The ages in one hospital may differ considerably from the ages in another. And the state of the cases on admission may differ very much in each hospital. These elements, no doubt, affect considerably the results of treatment, altogether apart from the sanitary state of hospitals.

In the next place accurate hospital statistics are much more rare than is generally imagined, and at the best they only give the mortality which has taken place *in* the hospitals, and take no cognizance of those cases which are discharged in a hopeless condition, to a much greater extent from some hospitals than from others.

We have known incurable cases discharged from one hospital, to which the deaths ought to have been accounted, and received into another hospital, to die there in a day or two after admission.

Again, the sanitary state of any hospital ought not to be inferred solely from the greater or less mortality. If the function of a hospital were to kill the sick, statistical comparisons of this nature would be admissible. As, however, its proper function is to restore the sick to health as speedily as possible, the elements which really give information as to whether this is done or not, are those which show the proportion of sick restored to health, and the average time which has been required for this object; a hospital which restored all its sick to health after an average of six months' treatment, could not be considered as by any means so healthy as a hospital which returned all its sick recovered in as many weeks. The proportion of recoveries, the proportion of deaths, and the average time in hospital must all be taken into account in discussions of this nature, as well as the character of the cases and the proportion of different ages among the sick.

Hospital mortality statistics* give little information on the point, because there are elements in existence of which such statistics take no cognizance. In one set of metropolitan hospitals, for example, I find the mortality about two and a-half per cent. upon the cases treated, while in other metropolitan hospitals the deaths reach from about twelve to sixteen per cent. To judge by the mortality in these cases would be most fallacious. Because in the first class of hospitals every ailment, however slight, constitutes a title to hospital admission, while, in the latter class of hospitals, special diseases only, at all times accompanied by a high rate of mortality, are admitted. Hence the duration of the cases admitted, and the general course and aspect of disease afford important criteria whereby to judge of the healthiness or unhealthiness of any hospital in

---

* In Paris an Annual Report of the Hospitals ('Compte Moral Administratif') is published. But the only useful statistical information to be gleaned from it is the number of *sous* each patient has cost. For, although it gives the numbers of adults, male and female, and of children who have been admitted, and who have died during the year, yet this in itself tells little.

If the Hospitals of London and of Paris would give us the information contained under the eight following heads, so important would be the knowledge thereby conveyed, that it would be worth while to go back for many years to construct such tables, and to continue the same forms hereafter.

1. The numbers admitted for each decennial period of age for each sex per annum.
2. The numbers, similarly arranged, remaining in hospital at the end of the preceding year.
3. The numbers dead for each sex at each decennial period of age per annum.
4. The numbers discharged cured similarly arranged per annum.
5. The numbers discharged incurable similarly arranged per annum.
6. The numbers remaining in hospital at the end of the current year similarly arranged.
7. The *diseases* remaining, admitted, died, cured, discharged incurable, and remained, arranged for each sex and each decennial period of age per annum.
8. The duration of cases, similarly arranged.

addition to that afforded by the mortality statistics. Besides, careful observers are now generally convinced that the origin and spread of fever in a hospital, or the appearance and spread of hospital gangrene, erysipelas and pyæmia generally, are much better tests of the defective sanitary state of a hospital than its mortality returns. But I would go further, and state that to the experienced eye of a careful observing nurse, the daily, I had almost said hourly, changes which take place in patients, and which changes rarely come under the cognizance of the periodical medical visitor, afford a still more important class of data, from which to judge of the general adaptation of a hospital for the reception and treatment of sick. One insensibly allies together restlessness, languor, feverishness, and general *malaise,* with closeness of wards, defective ventilation, defective structure, bad architectural and administrative arrangements, until it is impossible to resist the conviction that the sick are suffering from something quite other than the disease inscribed on their bed-ticket—and the inquiry insensibly arises in the mind, what can be the cause? To this query many years' experience of hospitals in various countries and climates enables me to answer explicitly as the result of my own observation, that, even admitting to the full extent the great value of the hospital improvements of recent years, a vast deal of the suffering, and some at least of the mortality in these establishments, is avoidable.

What, then, are those defects to which such results are to be attributed?

I should state at once that to original defects in the sites and plans of hospitals, and to deficient ventilation and overcrowding accompanying such defects, is to be attributed a large proportion of the evil I have mentioned.

The facts flow almost of necessity from ascertained sanitary experience. But it is not often, excepting perhaps in the case of intelligent house surgeons, that the whole process whereby the sick, who ought to have had rapid recoveries, are retained week after week, or perhaps month after month, in hospital, is continuously observed. I have known a case of slight fever received into hospital, the fever pass off in less than a week, and yet the patient, from the foul state of the wards, not restored to health at the end of eight weeks.

The defects to which such occurrences are mainly to be attributed are four:—

1. The agglomeration of a large number of sick under the same roof.
2. Deficiency of space.
3. Deficiency of ventilation.
4. Deficiency of light.

These are the four radical defects in hospital construction.

But on the very threshold of the subject we shall probably be told that not to these defects, but to 'contagion' and 'infection,' is much of the unhealthy condition of some hospitals attributable, at least so far as concerns the occurrence of zymotic diseases. On the very threshold, therefore, we are obliged to make a digression in order to discuss the meaning of these two familiar words, and to lay these spectres which have terrified almost all ages and nations.

This is the more necessary, because on the exact influence exercised by these two presumed causes of hospital sickness and mortality depends to a great degree the possibility of our introducing efficient hospital attendance and nursing. Unfortunately both nurses* and medical men, as well as medical students, have died of zymotic diseases prevailing in hospitals. It is an all-important question to decide whether the propagation of such diseases is inevitable or preventible.

---

* To show the great importance of this point I give the following tables, kindly prepared by Dr. Farr, from returns furnished to me with the greatest readiness by fifteen of the metropolitan hospitals. Table I. gives the ages of living and dying among the nursing staff. Table II. gives the mortality from zymotic diseases, and the comparison between the nurses' mortality and the mortality of the female population of London.

TABLE I.—*Numbers and Ages of Matrons, Sisters and Nurses (Living and Dying) in Fifteen London Hospitals.*

(*Names of the Hospitals,—St. Mary's; St. George's; Westminster; Charing Cross; Middlesex; University College; Royal Free; King's College; St. Bartholomew's; London; Guy's; St. Thomas'; Small Pox; Fever; and Consumption.*)

LIVING (1858).

	Total of all Ages.	Total.		Specified Ages of the Living, March, 1858.											
		Ages Specified.	Ages *not* Specified.	Under 20.	20.	25.	30.	35.	40.	45.	50.	55.	60.	65.	70 and up.
Matrons, Sisters and Nurses	521	391	130	1	10	45	55	93	64	59	34	18	8	4	...
Matrons and Sisters ...	118	90	28	...	...	4	11	22	16	20	8	5	3	1	...
Nurses	403	301	102	1	10	41	44	71	48	39	26	13	5	3	...

DYING (1848—57).

	Total of all Ages.	Total.		Ages of the Dying.											
		Ages Specified.	Ages *not* Specified.	Under 20.	20.	25.	30.	35.	40.	45.	50.	55.	60.	65.	70 and up.
Matrons, Sisters and Nurses	79	79	...	...	...	4	11	8	18	8	10	7	6	2	5
* Matrons and Sisters (so distinguished) ...	19	19	...	...	...	...	2	1	4	...	3	2	1	1	5
Nurses	60	60	...	...	...	4	9	7	14	8	7	5	5	1	...

* In the returns of deaths, four Hospitals do not distinguish the Matrons and Sisters from the Nurses, and in this Table they are included with the Nurses.

If the former, then the whole question must be considered as to whether hospitals necessarily attended with results so fatal should exist at all. If the latter, then it is our duty to prevent their propagation.

The idea of 'contagion,' as explaining the spread of disease, appears

---

TABLE II.—*Table of the Mortality of Matrons, Sisters, and Nurses, at different Ages, in Fifteen London Hospitals, compared with the Mortality of the Female Population of London.*

Ages.	Matrons, Sisters, and Nurses (1848—57).		Female Population of London.	
	Annual Rate of Mortality to 1000 living at the respective Ages.			
	By *all* returned *Diseases.*	By *Zymotic* Diseases.	By *Zymotic* Diseases (1848—57).	By *all* returned *Diseases* (1848—54).
25 to 35	15.89	9.53	2.19	9.92
35 — 45	15.80	10.94	2.73	14.65
45 — 55	17.80	11.87	3.17	20.36
55 — 65	46.36	14.26	4.94	36.02

The fatal zymotic diseases included in this table are fever and cholera, and it will be seen that these two diseases occasioned nearly 50 per cent. of the total mortality among the nursing staff as against 16 per cent. among the London female population. This single fact is quite enough to prove the very great importance of hospital hygiène. The calculated total mortality is also very much higher among the nurses, even if we assume that the deaths in the returns are all the deaths due to hospital nursing, which is very doubtful. If we assume that the non-zymotic mortality among nurses ought to be the same as it is among the female population, and if to this we add the zymotic deaths among nurses, we find the total mortality among nurses to exceed the total mortality among the female population of the metropolis by about 40 per cent. The loss of a well-trained nurse by preventible disease is a greater loss than is that of a good soldier from the same cause. Money cannot replace either, but a good nurse is more difficult to find than a good soldier.

The data from which these tables have been deduced are imperfect, and it would be very desirable if in future all hospitals would keep a register of nurses. The following form would be one well calculated to give the required information. The subject is of additional importance in connexion with the proper working of a Superannuation Fund for nurses:—

Form of Register of Sisters and Nurses in Hospital. Commenced from January 1, 1859.

No.	Name.	Age when first Appointed.	Date.		State of Health on Leaving the Service: and, if Ill, the Disease, its Duration, and probable Cause.	Date of Death, Cause of Death, and fatal Disease.
			Of Appointment.	Dismissal, Resignation, or Superannuation.		
1	Jane Jones. (Sister.)	25	June 6, 1858.	Resigned August 6, 1858.	In good Health.	
2	Mary Evans. (Nurse).	29	April 2, 1848.			July 7, 1854, Typhoid Fever, after 11 days' illness.

to have been adopted at a time when, from the neglect of sanitary arrangements, epidemics attacked whole masses of people, and when men had ceased to consider that nature had any laws for her guidance. Beginning with the poets and historians, the word finally made its way into medical nomenclature,* where it has remained ever since, affording to certain classes of minds, chiefly in the southern and less educated parts of Europe, a satisfactory explanation for pestilence and an adequate excuse for non-exertion to prevent its recurrence.

And now, what does 'contagion' mean? It implies the communication of disease from person to person by *contact.* It pre-supposes the existence of certain germs like the sporules of fungi, which can be bottled up and conveyed any distance attached to clothing, to merchandize, especially to woollen stuffs, for which it is supposed to have a particular affection, and to feathers, which of all articles it especially loves—so much so, that, according to quarantine laws, a live goose may be safely introduced from a plague country; but if it happen to be eaten on the voyage, its feathers cannot be admitted without danger to the entire community. There is no end to the absurdities connected with this doctrine. Suffice it to say, that in the ordinary sense of the word, there is no proof, such as would be admitted in any scientific inquiry, that there is any such thing as 'contagion.'

There are two or three diseases in which there is a specific virus,

* The history of the doctrine of 'Contagion' is given by Dr. Adams in his very learned translation of the works of Paulus Ægineta, Vol. 1, p. 284—(Sydenham Society). He says, in his comment, 'the earlier ancient authors appear to have entertained no suspicions of contagion as a cause of febrile or of other complaints.

'The works of the fathers of history, and of medicine, have likewise been ransacked in vain for any traces of the doctrine of contagion.'

Thucydides, and after him several of the Latin poets describe the plague of Athens, which appears to have been a form of Dysentery, as communicable from person to person. The later Greek historians contain allusions to the infectious nature of certain diseases; but Procopius, though cognizant of one of the greatest pestilences on record, was a non-contagionist.

Virgil's allusions to contagious diseases among cattle will be found in Ecl. I. Georg. III., 464.

Aretæus appears to be the first medical author who believed in contagion. Galen seems to have held the doctrine of infection. Of the later Greek and Arabian medical writers, some were contagionists, and others make no allusion to the subject. Dr. Adams states, in regard to plague, a disease which, in later times, has been considered as the very type of all 'contagious' pestilences, 'The result of our investigations into the opinions of the ancients on this subject leads us to the conclusion that all, or at least the most intelligent of the medical authorities, held that the plague was communicated not by any specific virus, but in consequence of the atmosphere around the sick being contaminated with putrid effluvia.'

The obvious practical result of this view of infection is, that abundance of pure air will prevent infection. All my own hospital experience confirms this conclusion. If infection exists, it is preventible. If it exists, it is the result of carelessness, or of ignorance. 'Contagion,' as a doctrine, on which distinct practical proceedings have been taken, appears to be of very modern invention; but it has been not the less injurious to civilization and humanity, from the loss of life which has from time to time followed from the practices which it inculcates, and from the immense tax which it has entailed upon commerce.

which can be seen, tasted, smelt, and analysed, and which in certain constitutions propagates the original disease by inoculation—such as small-pox, cow-pox, &c. But these are not 'contagions' in the sense supposed.*

The word 'infection,' which is often confounded with 'contagion,' expresses a fact, and does not involve a hypothesis. But just as there is no such thing as 'contagion,' there is no such thing as *inevitable* 'infection.' Infection acts through the air. Poison the air breathed by individuals and there is infection. Shut up 150 healthy people in a Black-hole of Calcutta, and in twenty-four hours an infection is produced so intense that it will, in that time, have destroyed nearly the whole of the inmates. Sick people are more susceptible than healthy people; and if they be shut up without sufficient space and sufficient fresh air, there will be produced not only fever, but erysipelas, pyæmia, and the usual tribe of hospital-generated epidemic diseases.

Again, if we have a fever hospital with over-crowded, badly-ventilated wards, we are quite certain to have the air become so infected as to poison the blood not only of the sick, so as to increase their mortality, but also of the medical attendants and nurses, so that they also shall become subjects of fever.

It will be seen at a glance, that in every such case and in every such example, the 'infection' is not inevitable, but simply the result of carelessness and ignorance. As soon as this practical view of the subject is admitted and acted upon, we shall cease to hear of hospital contagions.

In certain hospitals it has been the custom to set apart wards for what are called 'infectious' diseases, but in reality there ought to be no diseases so considered. With proper sanitary precautions, diseases reputed to be the most 'infectious' may be treated in wards among other sick without any danger. Without proper sanitary arrangements, a number of healthy people may be congregated together so as to become subject to the worst horrors of 'infection.'

No stronger condemnation of any hospital or ward could be pronounced than the simple fact that any zymotic disease has originated in it, or that such diseases have attacked other patients than those brought in with them. And there can be no stronger condemnation of any town than the outbreak of fatal epidemics in it. Infection, and incapable management, or bad construction, are in hospitals as well as in towns, convertible terms.

It was necessary to say thus much to show to what hospital diseases are *not* necessarily due. To the following defects in site, construction, and management, as we think, they are mainly to be attributed.

---

* Curiously enough, these directly communicable diseases were excluded from the operation of general quarantine law by the International Quarantine Conference of Paris, 1851, which restricted the objects of quarantine to plague, yellow fever, and cholera, while it gave a logical *coup de grace* to the 'contagion' hypothesis by abolishing the 'suspected bill of health.'

1. *The agglomeration of a large number of sick under one roof.*

It is a well-established fact that, other things being equal, the amount of sickness and mortality on different areas bears a ratio to the degree of density of the population.

Why should undue agglomeration of sick be any exception to this law? Is it not rather to be expected that, the constitutions of sick people being more susceptible than those of healthy people, they should suffer more from this cause?

But if anything were wanting in confirmation of this fact, it would be the enormous mortality in the hospitals which contained perhaps the largest number of sick ever at one time under the same roof, viz., those at Scutari. The largest of these too famous hospitals had at one time 2500 sick and wounded under its roof, and it has happened that of Scutari patients two out of every five have died. In the hospital tents of the Crimea, although the sick were almost without shelter, without blankets, without proper food or medicines, the mortality was not above one-half what it was at Scutari. Nor was it even so high as this in the small Balaclava General Hospital, while in the huts of the Castle Hospital, on the heights above Balaclava, at a subsequent period, the mortality among the wounded did not reach three per cent. It is not to this, however, that we appeal, as the only proof of the danger of surface over-crowding. It is to the fact of 80 cases of hospital gangrene having been recorded during one month at Scutari (and many, many more, passed unrecorded); to the fact that, out of 44 secondary amputations of the lower extremities consecutively performed, 36 died; and to the cases of fever which broke out in the hospital, not by tens but by hundreds.

All experience tells the same tale, both among sick and well. Men will have a high rate of mortality in large barracks, a low one in separate huts, even with a much less amount of cubic space.*

The example which France and Belgium have lately set us of separating their hospitals into a number of distinct pavilions, containing generally not more than 100 sick each, should be elsewhere imitated. It may be useful, by way of illustrating good and bad hospital structure, to annex plans of the newest civil and military hospitals constructed in Paris, in contrast with plans of the newest civil and military hospitals constructed in England.

The Lariboisière as a civil hospital, the Vincennes as a military one, exhibit the latest and the best specimens of hospital construction in Paris.

King's College as a civil hospital, Netley as a military one, are among the latest—we would we could say the best—plans of hospital construction in England.

The Lariboisière, as will be seen from the plan, contains 600 beds, under six different roofs.

---

* It must never be forgotten that, during the last six months of our occupation in the Crimea, the death-rate among our men, barracked in huts, was only two-thirds of what it is among the men in barracks at home.

In the Vincennes plan the pavilions are end to end, two and a-half in each wing, and contain about 600 beds in four pavilions and two half-pavilions.

Netley Hospital is to contain 1000 sick and invalids, under two roofs.

2. *Deficiency of Space.*—Wherever cubic space is deficient, ventilation is bad. Cubic space and ventilation will therefore go hand in hand. The law holds good with regard to hospitals, barracks, and all inhabited places. Deficiency of cubic space is confounded by unskilful sanitary statisticians with surface over-crowding in towns, although the things are quite different, and lead to different results. In a recent paper it has been argued that because the statistics of disease in towns of different densities do not show so large a proportionate mortality from consumption as takes place in the army, *therefore* the allegation that the army mortality is caused by overcrowding and bad ventilation is incorrect. We happen to know that deficient external ventilation and over-crowding in barracks, as regards cubic space, stand as follows:—

The cavalry barracks, as a whole, are the least over-crowded, and have the freest external movement of air. Next come the infantry; and the most crowded and the least ventilated externally are the Guards' barracks; so that the mortality from consumption which follows the same order of increase in the different arms augments with increase of crowding, and difficulty of ventilation.

If over-crowding or its concomitant, bad ventilation, among healthy people, generates disease, it does so to a far greater extent among the sick in hospitals. In civil hospitals the amount of cubic space varies between 600 and 2000 cubic feet per bed. In some military hospitals it is under 300; and from 700 to 800 appear to be considered a somewhat extravagant allowance. The army regulation as to cubic space in hospitals IS over-crowding. At Scutari, at one time, not even half the regulation-space was given; and the great over-crowding consequent thereupon was one element in the disastrous result which followed. Any one in the habit of examining hospitals with different relative amounts of cubic space cannot fail to have been struck with the very different appearance of the sick, and with the different state of the ward atmosphere. It is impossible to ventilate a ward in a brick or stone hospital by natural means, when the cubic space is less than a certain amount. Crowded wards are, in fact, offensive, with all the windows open.

In the country less cubic space is essential than in towns. In detached huts or pavilions, especially if they be but one story high, less cubic space is necessary than where numbers are massed together.

Under all circumstances, however, the progress of the cases (in solidly-built hospitals) will betray any curtailment of space much below 1500 cubic feet. In Paris 1700, and in London 2000 and even 2500 cubic feet are now thought advisable.

The master of some large works in London lately mentioned the following fact:—He was in the habit of sending those of his workmen who met with accidents to two different metropolitan hospitals. In one they recovered quickly: in the other they were frequently attacked with erysipelas, and some cases were fatal. On inquiry it appeared that in the former hospital a larger amount of cubic space was allowed than in the latter, which is also so deficient in external ventilation and in construction, that nothing but artificial ventilation could effectively change its atmosphere.

It is no less important to have a sufficient surface-area between the adjoining and the opposite beds. Piling cubic space above the patient is not all that is wanted. In the lofty corridors of Scutari I have seen two long rows of opposite beds with scarcely three feet from foot to foot. Certainly it cannot be thought too much, under any circumstances, to give to each bed a territory to itself of at least eight feet wide by twelve feet long.

3. *Deficiency of Ventilation.*—The want of fresh air may be detected in the appearance of patients sooner than any other want. No care or luxury will compensate indeed for its absence. Unless the air *within* the ward can be kept as fresh as it is *without*, the patients had better be away. Except in a few cases well known to physicians the danger of admitting fresh air directly is very much exaggerated. Patients in bed are not peculiarly inclined to catch cold,* and in England, where fuel is cheap, somebody is indeed to blame, if the ward cannot be kept warm enough, and if the patients cannot have bed-clothing enough, for as much air to be admitted from without as suffices to keep the ward fresh. *No* artificial ventilation will do this. Although in badly-constructed hospitals, or in countries where fuel is dear, and the winter very cold, artificial ventilation may be necessary, it never can compensate for the want of the open window. The ward is never fresh, and in the best hospitals at Paris, artificially ventilated, it will be found that, till the windows are opened in the morning, the air is close. Natural ventilation, or that by open windows and open fire-places, is the only efficient means for procuring the life-spring of the sick—fresh air. But to obtain this the ward should be at least sixteen feet high, and the distance between the opposite windows not more than thirty feet. The amount of fresh air required for ventilation has been hitherto very much underrated, because it has been assumed that the quantity of carbonic acid produced during respiration was the chief noxious gas to be carried off. The total amount of this gas produced by an adult in twenty-four hours is about 40,000 cubic inches, which, in a barrack-room, say, containing sixteen men, would give 370 cubic feet *per diem.* Allowing eight hours for the night occupation of such a room, when the doors and windows may be supposed to be shut, the product of carbonic acid would be 123 cubic feet, or

* 'Catching cold' in bed follows the same law as 'catching cold' when up. If the atmosphere is foul, and the lungs and skin cannot therefore relieve the system, then a draught upon the patient may give him cold. But this is the fault of the foul air, not of the fresh.

about fifteen and a-half cubic feet per hour. This large quantity, if not speedily carried away, would undoubtedly be injurious to health; but there are other gaseous poisons produced with the carbonic acid which have still greater power to injure. Every adult exhales by the lungs and skin forty-eight ounces, or three pints of water in twenty-four hours. Sixteen men in a room would therefore exhale in eight hours sixteen pints of water, and 123 cubic feet of carbonic acid into the atmosphere of the room. With the watery vapour there is also exhaled a large quantity of organic matter, ready to enter into the putrefactive condition. This is especially the case during the hours of sleep, and as it is a vital law that all excretions are injurious to health if reintroduced into the system, it is easy to understand how the breathing of damp foul air of this kind, and the consequent re-introduction of excrementitious matter into the blood through the function of respiration will tend to produce disease.

If this be so for the well, how much more will it be so for the sick? —for the sick, the exhalations from whom are always highly morbid and dangerous, as they are one of nature's methods of eliminating noxious matter from the body, in order that it may recover health.

One would think that the first and last idea in constructing hospitals would be to contrive such means of ventilation as would be perpetually and instantly carrying off these morbid emanations. One would think that it would be the first thing taught to the attendants to manage such means of ventilation. Often, however, it is *not even* the *last* thing taught to them.

A much larger mass of air is required to dilute and carry away these emanations than is generally supposed, and the whole art of ventilation resolves itself into applying in any specific case the best method of renewing the air sufficiently without producing draughts, or occasioning excessive varieties in temperature. Trifling varieties are rather beneficial than otherwise in most cases. A cooler atmosphere at night acts like a tonic.

4. *Deficiency of Light.*—What is the proportionate influence of the four defects enumerated in delaying recovery I am not competent to determine.

Second only to fresh air, however, I should be inclined to rank light in importance for the sick. Direct sunlight, not only daylight, is necessary for speedy recovery, except, perhaps, in certain ophthalmic and a small number of other cases. Instances could be given, almost endless, where, in dark wards or in wards with a northern aspect, even when thoroughly warmed, or in wards with borrowed light, even when thoroughly ventilated, the sick could not by any means be made speedily to recover. The effect of light on health and disease has been ably discussed in an article on light in the August number, 1858, of the 'North British Review.' Its importance has been long recognised in the medical profession, as may be learned from the writings of Sir Andrew Wylie, Dr. Milne Edwards, and Mr. Ward. Dark barrack-rooms, and barrack-rooms with northern aspects, will furnish a larger amount of sickness than light and sunny rooms.

Among kindred effects of light I may mention, from experience, as quite perceptible in promoting recovery, the being able to see out of a window, instead of looking against a dead wall; the bright colours of flowers; the being able to read in bed by the light of a window close to the bed-head. It is generally said that the effect is upon the mind. Perhaps so; but it is no less so upon the body on that account.

All hospital buildings in this climate should be erected so that as great a surface as possible should receive direct sunlight—a rule which has been observed in several of our best hospitals, but, I am sorry to say, passed over in some of those most recently-constructed. Window-blinds can always moderate the light of a light ward; but the gloom of a dark ward is irremediable.

The axis of a ward should be as nearly as possible north and south; the windows on both sides, so that the sun shall shine in (from the time he rises till the time he sets) at one side or the other. There should be a window to at least every two beds, as is the case now in our best hospitals. Some foreign hospitals, in countries where the light is far more intense than in England, give one window to every bed. The window-space should be one-third of the wall-space. The windows should reach from two or three feet of the floor to one foot of the ceiling. The escape of heat may be diminished by plate or double glass. But while we *can* generate warmth, we cannot generate daylight, or the purifying and curative effect of the sun's rays.

PART II.

Considering, then, that the conditions essential to the health of hospitals are principally these—

1. Fresh Air. 2. Light. 3. Ample Space. 4. Subdivision of Sick into Separate Buildings or Pavilions—let us examine the causes in the usual ward construction which prevent us from obtaining these and other necessary conditions. The principal causes are as follow, viz.:—

1. Defective Means of Natural Ventilation and Warming.
2. Defective Height of Wards.
3. Excessive Width of Wards between the Opposite Windows.
4. Arranging the Beds along the Dead Walls.
5. Having more than two Rows of Beds between the Opposite Windows.
6. Having Windows only on one Side, or having a closed Corridor connecting the Wards.
7. Using Absorbent Materials for Walls and Ceilings, and Washing Floors of hospitals.
8. Defective Condition of Waterclosets.
9. Defective Ward Furniture.

10. Defective Accommodation for Nursing and Discipline.
11. Defective Hospital Kitchens.
12. Defective Hospital Laundries.
13. Selection of Bad Sites and Bad Local Climates for Hospitals.
14. Erecting Hospitals in Towns.
15. Defects of Sewerage.
16. Construction of Hospitals without Free Circulation of External Air.

1. *Defective Means of Ventilation and Warming.*—When the question of ventilation first assumed a practical shape in this country, it was supposed that 600 cubic feet of air per hour were sufficient for a healthy adult, in a room where a number of people are congregated together. Subsequent experience, however, has shown that this is by no means enough. As much as 1000 cubic feet have been found insufficient to keep the air free from closeness and smell; and it is highly probable that the actual quantity required will ultimately be found to be at least 1500 cubic feet per hour per man.

In sick wards we have more positive experience as to the quantity of air required to keep them sweet and healthy. It has been found in certain Parisian hospitals, in which the ventilating arrangements were deficient, that pyæmia and hospital gangrene had appeared among the patients. These diseases disappeared, on the introduction of ventilating arrangements, whereby 2500 cubic feet of air per bed per hour were supplied to the wards. Notwithstanding this large quantity, however, the ward-atmosphere was found not to be sufficiently pure. In other wards the quantity of air was increased to as much as 4000 or 5000 cubic feet per bed per hour—an amount which keeps the wards perfectly sweet. But again we say, do not trust to artificial means; without natural ventilation the air will never be *fresh*.

In this country, have no other than the open fireplace. It is the safest warmer and ventilator. Heated air from metal surfaces is especially to be avoided. It seems likely that we shall soon be enabled to have open fireplaces in the middle of wards, the draught being carried under the floor. It is obvious that fireplaces in the side walls are in the wrong place. There is great loss and unequal distribution of heat in consequence.

2. *Defective Height of Wards.*—It is not possible to ventilate sufficiently a ward of ten or twelve feet high. And again, it is not possible to ventilate a ward where there is a great height above the windows. A ward of thirty beds can be well ventilated with a height of about sixteen or seventeen feet, provided the windows reach to within one foot of the ceiling. Otherwise, the top of the ward becomes a reservoir for foul air.

3. *Too Great Width of Wards between the Opposite Windows.*—It does not appear as if the air could be thoroughly changed, if a distance of more than thirty feet intervenes between the opposite windows: if, in other words, the ward is more than thirty feet wide.

This is the true starting-point from which to determine the size of your ward, and the number of beds you will have in it. If you make your length too great in proportion to this width, your ward becomes a tunnel—a form fatal to good ventilation. This was the case with the great corridor wards at Scutari.

If, on the other hand, you make your wards too short in proportion to this width, you multiply corners in a greater ratio than you multiply sick. And direct experiment has shown that the movement of the air in the centre of a ward is three or four times as great as it is at the corners. The movement of the air in a hospital ward should always be slightly perceptible over the face and hands, and yet there should be no draughts.

4. *Arranging the Beds along the Dead Walls.*—This deprives the patient of the amount of light and air necessary to his recovery, and has, besides, the disadvantage that when the windows are opened the effluvia must blow over all the intervening beds before escaping. This arrangement is to be seen at Portsmouth Military Hospital, Chatham Garrison Hospital, in the new part of the Edinburgh Infirmary, and is proposed at Netley Hospital.

5. *Having more than Two Rows of Beds between the Windows.*—In the double wards, or wards back to back, of the new part of Guy's, of King's College, and of the Fever Hospital, this arrangement is seen. It is objectionable on every account. These double wards are from twelve to nearly twenty feet wider than they ought to be between the opposite windows for thorough ventilation. The partition down the middle with apertures makes matters rather worse; complaint has been made that it beats down the draught on the heads of the inner rows of patients. It also prevents the head nurse from having that view of her whole ward at once, which she ought to have for proper care of it. The only hospital in which this arrangement of four rows of beds could be comparatively unobjectionable, would be in a one-storied hut hospital, ventilated through the ceiling, like that of Dr. Parkes, at Renkioi.* But his were magnificent huts, and the partition was little more than a bulkhead. In the ordinary huts of the Sardinian camp-hospitals at Balaclava I have seen this arrangement produce pernicious effects.

6. *Having Windows only on one Side, or having a closed Corridor connecting the Wards.*—As it is a necessity of hospital construction that every ward ought to have direct communication with the external air by means of a sufficient number of windows on its opposite sides, it follows that to have a dead wall on one side, or to cover one of the sides by a corridor, is directly to interfere with the natural ventilation of the ward. To join all the ward doors and windows on one side by means of a corridor is much more objectionable than even

* On the Dardanelles during the Crimean War.

to have a dead wall, because the foul air of all the wards must necessarily pass into the corridor; and hence, without extraordinary precautions, such as are not usually nor likely to be bestowed on such matters, these corridors are the certain means of engendering a hospital atmosphere. If any one had wished to see the corridor plan in all its horrors, Scutari would have shown them to him on a colossal scale. But the evils connected with corridors may be seen on a smaller scale in almost every hospital in London, and Netley also is to have its corridor.

This country is much indebted to Mr. Roberton, of Manchester, to the Medical Staff of the Middlesex Hospital, and to the Army Sanitary Commission for their advocacy of the pavilion system of hospital construction, in opposition to the corridor system, as also for their enlightened labours in the cause of good hospital construction generally.

7. *Using Absorbent Materials for Floors, Walls, and Ceilings of Hospitals, and Washing Floors.*—The amount of organic matter given off by respiration and in other ways from the sick is such that the floors, walls, and ceilings of hospital wards—if not of impervious materials, become dangerous absorbents.

The boards are in time saturated with organic matter, and only require moisture to give off noxious effluvia. When the floors are being washed, the smell of something quite other than soap and water is perfectly perceptible, and there cannot be a doubt that washing floors is one cause of erysipelas, &c., in some hospitals.

In Scutari, where the wards were overcrowded, the cases offensive, and the floors ill-laid, rotten and dirty, the accumulated saturations of weeks and months were such that the floors could not be scoured without poisoning the patients.

There is no remedy for this but filling up the grain of the wood (which ought to be oak) with bees-wax and turpentine, like the French *parquet*, or oiling and *lackering*, *i. e.*, saturating the floor with linseed-oil, and then rubbing it over with a peculiar *laque* varnish, and polishing it so as to resemble French polish, like the Berlin hospital floors. Both processes render the floor non-absorbent—both processes do away with the necessity of scouring altogether. The French floor *stands* the most wear and tear, but must be cleaned by a *frotteur*, which cleaning is more laborious than scrubbing, and does not remove the dust. The Prussian floor requires re-preparing every three years. But the wet and dry rubbing, or process of cleaning is far less laborious than either *frottage* or scrubbing, and completely removes the dust, and freshens the ward in the morning. By either process the sick would gain much in England. The Berlin flooring is by no means perfect, on account of this deficient durability of surface, and might be improved.

As to the walls and ceilings of wards, plaster, or brick whitewashed, are equally objectionable. Pure, white, polished, non-absorbent cement is the only material fit for hospital walls. If any one has inhabited the wards of War Hospitals, after several weeks or months of constant occupation by sick and wounded, where little or no attempt had been made to lime-wash the uneven dirty plaster-walls, saturated with organic matter, he will not wonder at the stress which is here laid upon the importance of impervious walls.

8. *Defective Condition of Waterclosets.*—It is hardly necessary to say more than this. There can be no safety for the sick if any but waterclosets of the best construction are used, as also if they are not built *externally* to the main building, and cut off by a lobby, separately lighted and ventilated, from the ward. The same thing may be said of sinks. I have known outbreaks of fever even among the healthy from an ill-constructed and ill-placed sink in this country.

The smell of latrines, which are not waterclosets, as used in French hospitals, although externally built, is quite perceptible at the end of the ward nearest to them.

9. *Defective Ward Furniture.*—Hospital bedsteads should always be of iron, the rest of the furniture of oak. Hair is the only material yet discovered fit for hospital mattresses. It is not hard nor cold. It is easily washed. It does not retain miasma. Straw has the advantage of being easily renewed, but it is not desirable. It is too hard and too cold not to render necessary the use of a blanket *under* the patient, which use is likely to encourage bed-sores. I speak from actual experience of the fatal effect of using the paillasse with patients much reduced. It may lower their vital energy beyond repair.

For all eating, drinking, and washing vessels, and for other utensils, the use of glass or earthenware is superior to that of tin or any other metal, on account of its greater cleanliness. Notwithstanding the greater amount of breakage and of expense, glass or earthenware is therefore best wherever possible. Some kinds of tin vessels cannot by any amount of cleaning be freed from an unclean smell.

10. *Defective Accommodation for Nursing and Discipline.*—Simplicity of construction in hospitals is essential to discipline. Effectual and easy supervision is essential to proper care and nursing.

Every unneeded closet, scullery, sink, lobby, and staircase represents both a place which must be cleaned, which must take hands and time to clean, and a hiding or skulking place for patients or servants disposed to do wrong. And of such no hospital will ever be free. Every five minutes wasted upon cleaning what had better not have been there to be cleaned, is something taken from and lost by the sick.

In considering the pavilion plan to be in future received as the sanitary necessity for hospital construction, we must look upon it as susceptible of many modifications. In deciding which of these shall be adopted, there are four essentials to be considered as regards the head of nursing and discipline. 1. Economy of attendance. 2. Ease of supervision. 3. Convenience as to number of sick in the same ward and on the same floor, so as to save extra attendants and unnecessary waste of time and strength on the stairs. 4. Efficiency as to accommodation for nurses so as to overlook their wards.

First. *Economy as to attendance.*—I would rather not enumerate the instances where I have seen that, often from the most various causes, one result arises, viz., that more time and care are given to passages, stairs, &c. &c., than to the sick. Extreme simplicity of construction and of detail is essential to obviate this. A convenient arrangement of lifts, and the laying of hot and cold water all over the building economize attendance—certainly as much as one attendant to every thirty sick.

Secondly. *Ease of Supervision.*—The system of scouts, watch, alarm, is well understood in many wards where patients would be puzzled to give the things names. Some patients will know both things and names. Attendants require inspection as well as patients. Whatever system of hospital construction is adopted should provide for easy supervision at unexpected times. The Vincennes plan is better adapted for this than the Lariboisière plan, inasmuch as there is a greater number of patients on the same level, and stairs are spared.

Third and Fourth. *Distribution of Sick in convenient numbers for attendance, and Position of Nurses' Rooms.*—Four wards of ten patients each, taking the average of patients as in London, cannot be efficiently overlooked by one head nurse. Forty patients in one ward can be fully overlooked by one head nurse. She ought to have her room so placed that she can command her whole ward, day and night, from a window looking into the ward. This cannot be the case if she has four wards. If she has two, they ought to be built end to end, with her room placed between and looking into both wards.

Four wards of ten patients each cannot be attended by one night nurse, taking the average of London cases. Forty patients in one ward can be fully attended by one night nurse.

Small wards are indeed objectionable in working a hospital.

If we are to be guided, however, by the results of recent experience in hospital building, we shall probably come to the conclusion that, taking sanitary and administrative reasons together, thirty-two patients is a good ward-unit.

Let us see what we do in our military hospitals at home. The first thing that will strike any one in most of our regimental hospitals is the extraordinary number of wards, and of holes and corners in comparison with the number of sick. In a hospital for a battalion 500 or 600 strong, you find eight or ten little bed-rooms, miscalled wards, a little kitchen, everything, in fact, on a little scale, like a collapsed French hospital. How much more sensible would it be to have one, or at most two large wards for thirty sick each, with a small 'casualty' ward! How much less the expense of erection and administration, how much easier the discipline and oversight, how much better the ventilation!

To return to large general hospitals. These 'casualty' wards, as they are called, for noisy or offensive cases are much better placed apart, with a completely appointed staff of their own, than attached one small ward to each larger one. Patients requiring much attention, whose condition fits them the most for the small wards, cannot be put there, because either they are more or less neglected or they unduly monopolize the service of the ward attendants. If convalescent patients are put into them, they are comparatively removed from inspection, and often play tricks there. If separate 'casualty' wards are provided as they ought to be, the small ward (often seen in French hospitals), at the end of the larger ward, is only an incubus.

11. *Defective Hospital Kitchens.*—Two facts every careful observer can establish from experience.

1. The necessity for variety in food, as an essential element of health, owing to the number of materials required to preserve the human frame. In sickness it is still more important, because, the frame being in a morbid state, it is scarcely possible to prescribe beforehand with certainty what it will be able to digest and assimilate. The so-called 'fancies' of disease are in many cases valuable indications.

2. The importance of cooking so as to secure the greatest digestibility and the greatest economy in nutritive value of food.

Yet so little was either of these elements of health understood in the late Crimean war, so little is either understood up to this hour in the diets, rations, and cooking of either sick or well in the army, that we still see the everlasting sameness of ration, the eternal boiled meat of the 'full,' 'half,' and 'low' diet of the hospital kitchen. As the present Quartermaster-General says, 'the men live upon boiled meat for twenty-one years.'

In the war hospitals of the East it was practically learnt, though never theoretically acknowledged, that in order to make the patient eat at all, he must not be fed on this hard boiled never-varying meat of 'full' or 'half' diet.

Hence the wasteful and violent expenditure of 'extras,' as they are called, which, in the hands of inexperienced surgeons, left to their own unassisted inventions, often laid them open to criticism.

But no attempt was made to compose a better or more varied diet, hardly any, till the arrival of Soyer, to improve the system of cooking.

Some few improvements have lately been made in the hospital kitchens of the army at home; what variety of cooking there is even now in the barracks is often done at the expense of, and by the men themselves. A proper system of hospital diets will ere long be introduced in the army hospitals.

I have often been surprised by the primitive kitchens of some of our civil hospitals, with which little variety of cooking is possible.

These things show how little diet and cooking are even yet thought of as sanitary and curative agents. There still exists a confusion of ideas about 'spoiling' the sick, about 'too much indulgence' of the patients, and even yet comparatively little is practically known as to what is, and what is not, essential for restoration to the utmost vigour.

12. *Defective Hospital Laundries.*—It is hardly necessary to go back to the time in the Crimean war when in a Scutari hospital six shirts were washed in a month for a number of 2000 patients, which was constantly changing; when the number per man per month of all articles of all descriptions washed was less than three. The pestilential filth of that time is known now to all. But it is not so well known that even now there is scarcely an army hospital which has such a thing as a laundry. The bedding is generally washed by the barrack department; no one appears to know how. It is done by contract. And the body linen is generally washed, if such a term ought to be used, in a small wash-house, or lean-to shed, with or without a boiler, and without any means for drying, getting-up, or airing linen. The linen is taken out of the damp wash-house, possibly into the damp air, and there hung up for a longer or a shorter time; and if the 'orderly' be careful of his patients, he will complete the process by drying the linen, before it is put on, in front of the ward fire.

A great deal has been said about the communication of 'infectious' disease, both in civil and military hospitals, from patients' linen to washerwomen. The usual conclusion arrived at on such occasions is that such and such a disease is 'very infectious;' *e.g.* I was lately told in a civil hospital that the washerwomen became infected with fever from the patients' linen. Have those who put forward this doctrine of inevitable 'infection' among washerwomen ever examined the process of washing, the appliances by which it is done, and the place where the women wash? If they will do so, they will very generally find a small, dark, wet, unventilated, and overcrowded little room or shed, in which there is hardly space to turn about—so full of steam loaded with organic matter that it is hardly possible to see across the room. Is it surprising that the linen is badly washed, that it is imperfectly dried, and that the washerwomen are poisoned by inhaling organic matter and foul air? An ordinary hospital wash-house is a very likely place indeed to contract disease in, but it supplies equal reason for demurring *in toto* to the doctrine that the occurrence is inevitable, or that the disease is to blame. Ignorance and mismanagement lie at the root of all such presumed cases of 'infection.' And it would better serve the cause of humanity if, instead of citing such facts—if they be facts—as illustrations that such and such a disease is infectious, people would reform these washing establishments and convert them into proper laundries, from which properly cleansed and prepared linen could be supplied to the sick, and in which the health of the servants could be preserved from injury.

Let laundries be constructed with sufficient area and cubic space for each washer, with abundance of water, with proper means of drainage, and of ventilation for removing the vapour, and with properly-constructed drying and ironing rooms, and we shall cease to hear of washerwomen 'catching' fever.

13. *Selection of Bad Sites and Bad Local Climates for Hospitals.*—As the object to be attained in hospital construction is to have pure dry air for the sick, it will be evident that this condition cannot be fulfilled if a damp climate be selected. It is a well-known fact, *e.g.*, that in the more damp localities of the south of England, certain classes of sick and of invalids linger, and do not recover their health. Again, retentive clay subsoils keep the air over entire districts of the country always more or less damp. And soils of this character should not be selected as sites for hospitals. Self-draining, gravelly, or sandy subsoils are best. River banks, estuary shores, valleys, marshy or muddy ground, ought to be avoided. It may seem superfluous to state that a hospital should not be built over an old graveyard, or on other ground charged with organic matter. Although hospitals are intended for the recovery of health, people are very apt to forget this, and to be guided in the selection of sites by other considerations—such as cheapness, convenience, and the like; whereas, the professed object in view being to secure the recovery of the sick in the shortest time, and to obtain the smallest mortality, that object should be distinctly kept in view as one which must take precedence of all others.

A doctrine has recently been promulgated in a Government Report, that we are only to consider what is best for the *majority* of the sick in a hospital. If we cannot do the best possible for *all* the sick, by all means let us leave the rest at home. In practice a hospital may be found only to benefit a majority, and to inflict suffering on the remainder. Let us use our intelligence to see whether we cannot have hospitals constructed so as to be of equal benefit to all.

14. *Erecting Hospitals in Towns.*—Nearly all that has been said under the last head, *mutatis mutandis*, may be repeated here. If the recovery of the sick is to be the object of hospitals, they will not be built in towns. If medical schools are the object, surely it is more instructive for students to watch the recovery from, rather than the lingering in, sickness. Twice the number of cases would be brought under their notice in a hospital in which the sick recovered in half the time necessary in another.

According to all analogy, the duration of cases, the chances against complete recovery, the rate of mortality, must be greater in town than in country hospitals.

Land in towns is too expensive for hospitals to be so built as to secure the conditions of ventilation and of light, and of spreading the inmates over a large surface-area, instead of piling them up three or four stories high—conditions now known to be essential to recovery.

15. *Defects in Drainage.*—Sewers may become cesspools of the most dangerous description, if improperly made and placed. At Scutari, if

the wind changed so as to blow up the open mouths of the sewers, such change was frequently marked by outbreaks of fever among the patients, and by relapses among the convalescents from fever. Where there are no means for externally ventilating the sewers, no means for cleansing or flushing them, and where the bottoms are rough and uneven, such occurrences cannot fail to take place. The emanations from the deposits in the sewers were blown back through the pipe-drains into the privies, and thence into the corridors and wards where the sick were lying. Where sewers pass close to or under occupied rooms, the walls or covers being defective, exhalations will infallibly escape into those rooms. Such could be distinctly perceived in Scutari hospitals, and cases of cholera distinctly traced to such a cause.

Not very long ago five fatal cases of fever occurred in rapid succession among the nurses in one of our civil hospitals, which were traced to a defective drain.

Where a main sewer is too large, as is the case at Netley Hospital, mischief may also ensue.

16. *Construction of Hospitals without Free Circulation of External Air.*—To build a hospital with one closed court with high walls, or what is worse, with two closed courts, is to stagnate the air even before it reaches the wards.

All closed corners stagnate the air, even where the building forms but three sides of a square, unless the wings are so short that they can hardly be called wings. The only safe plan is to leave the corners entirely open, as at Vincennes, where they are connected only by an arcade on the ground floor.

Even in the pavilion structure, unless the distance between the pavilions be double the height of the walls, the ventilation and light are seriously interfered with.

For this, among other reasons, two stories are better than three and one is preferable to two, provided it be erected upon an arched basement.

To build a hospital in the midst of a crowded neighbourhood of narrow streets and high houses, is to insure a stagnation of the air without, which no ventilation within, no cubic space, however ample, will be able to remedy.

I have here given the defects; few have had so sad or so large an experience of their results as I have had. I appeal to those who are wiser, and have more practical power than I have, for the remedies —to architects, to hospital committees, to civil and royal engineers, to medical officers, to officers of health, to all the men of science and benevolence of whom our country is so justly proud. It is hard that in a country where everything is done by a despotic Government, such advances in the sanitary construction of hospitals should have been made, and that our England, which ought to take the lead in everything good, should be left behind.

NOTE ON THE HOSPITAL PLANS.

THESE four plans, two English and two French, may be taken as representing the degree of constructive ability directed to the planning of hospitals in the two countries. The contrast presented by them is certainly very striking.

Compare, for example, the extreme simplicity of the plan of the Military Hospital at Vincennes with the great complication of that at Netley. The former consists of a centre and two wings, detached but connected by a corridor one story high running along the basement. The centre contains the offices, and the wards are in the wings. Each wing consists of two pavilions for sick soldiers, and one small pavilion for sick officers. The pavilions are completely cut off from each other by a large, specially ventilated staircase, carried above the roof. Each ward has a profusion of windows opposite each other, abundance of light and of ventilation, quite independent of the ventilation of the adjoining pavilions. The wards, moreover, run nearly north and south, and receive the sunlight freely throughout the day.

Netley Hospital, on the other hand, presents a perfect rabble of wards and offices, thrown together as if by accident. All the sick wards in each flat have their ventilation connected by a corridor running along and covering the whole south-west face of the building, as if designed to obstruct natural ventilation, to keep out sunlight, and to ensure the equal diffusion of a hospital atmosphere throughout the entire line of wards. It is true that a complicated system of ventilation is proposed to remove this latter defect, but there should have been no such defect to remove. It will also be seen, that the only ward windows opening to the external air are on the north-east and coldest side of the building.

Netley may be fairly described as a hospital without sufficient sunlight or natural ventilation.

The Vincennes Hospital has an obvious defect in the position of part of the administrative offices; but even in this respect it is better than Netley, while in all that pertains to the welfare of the sick it is very greatly superior.

Let us next compare the Civil Hospital Plans:—

One of these is that of an unfinished metropolitan hospital; the other is the plan of the noble Lariboisière at Paris. The English hospital plan presents an epitome of almost every defect in hospital construction. It is an involved Netley plan with sick in the corridors, for not only have the wards windows only on one side, but there are four rows of beds between the opposite windows. Moreover, the arrangement of wards, corridors, chapel, &c., seems as if intended to stagnate the air outside. All is complicated, and there is a want of that simplicity of plan which is essential to the free circulation of air without as well as within the sick wards.

Turn next to the Lariboisière. The corridor running round the central garden is only one story high, so as not to impede the free flow of the air and sunlight. The sick pavilions are all detached from each other. They are, in fact, separate hospitals, with only one hundred sick under one roof. Like the Vincennes pavilions, they have a profusion of windows, and abundant means of natural ventilation within as well as without. In the much brighter and hotter, as well as colder climate of Paris, a large proportion of the hospital wall is glass, and the sick are ranged between the windows, so that the ward effluvia can readily escape.

The English plans, on the contrary, show that in our duller and milder climate, in both senses, our hospital architects do their best to shut out our rare and imperfect sunlight, and to keep pure air out of the wards as much as possible, while they provide for the sick being so arranged that the effluvia must pass over a succession of beds before escaping.

Any one making even a cursory examination of these four plans can hardly fail to arrive at the conclusion that the French plans, with certain obvious defects, show a high appreciation of the importance of hospital hygiène; while the English plans, on the contrary, prove that we have hardly yet begun even to study this branch of knowledge.

ANSWERS TO WRITTEN QUESTIONS

ADDRESSED TO MISS NIGHTINGALE BY THE COMMISSIONERS APPOINTED TO INQUIRE INTO THE

Regulations affecting the Sanitary Condition of the Army,

Reprinted (with some alterations) from the Report of the Royal Commission.

I.—Sanitary State of the Army and Hospitals.

HAVE you, for several years, devoted attention to the organization of civil and military hospitals?

Yes, for thirteen years.

What British and foreign hospitals have you visited?

I have visited all the hospitals in London, Dublin, and Edinburgh, many county hospitals, some of the naval and military hospitals in England; all the hospitals in Paris, and studied with the 'sœurs de charité;' the Institution of protestant deaconesses at Kaiserswerth, on the Rhine, where I was twice in training as a nurse; the hospitals at Berlin, and many others in Germany, at Lyons, Rome, Alexandria, Constantinople, Brussels; also the war hospitals of the French and Sardinians.

When were you sent out to the British war hospitals at Constantinople?

We arrived at Constantinople on November 4, 1854, the eve of the Battle of Inkermann.

What hospitals did you find occupied there by the British?

Two large buildings on the Asiatic side, near Scutari, viz., a Turkish barrack and a Turkish military general hospital, both of which had been given over by the Turkish government for the use of the British troops.

How many patients did they contain at that date?

The Barrack contained 1500, the General hospital 800 patients, total 2300.

By how many nurses and ladies were you accompanied?

By 20 nurses, 8 Anglican 'sisters,' 10 nuns, and 1 other lady.

Where did you take up your residence?

We were quartered the same evening in the Barrack hospital, and two months afterwards, a reinforcement of 47 nurses and ladies having been received from England, we had additional quarters assigned us in the general hospital, and later at Koulali.

How long did you reside in those hospitals?

Till the evacuation of Turkey by the British, July 28, 1856.

Did you visit the hospitals in the Crimea?

Three times. I *visited* the regimental hospitals; I remained in the Crimean general hospitals about six months.

For what purpose did you visit the Crimea?

Each time to place or to reinforce nurses in the Crimean general hospitals. The first time, for the General and Castle hospitals at Balaclava; the second, for the Castle and Monastery hospitals; the third, for the two land transport general hospitals, for which nurses had been 'required' by the principal medical officer of the army in the east.

Was that purpose effected?

Yes; although difficulties sometimes interrupted the work.

To ascertain the efficiency of the sanitary or medical organization of the army, it should be tested by its results in peace and in war?

Certainly, in both.

What tests under those two conditions exist, and are available for our instruction, particularly in reference to the state of war?

The barrack, and the military hospital exist at home and in the colonies as tests of our sanitary condition in peace; and the histories of the Peninsular war, of Walcheren, and of the late Crimean expedition, exist as tests of our sanitary condition in the state of war.

Is it necessary that you should refer at all to the hospitals of Scutari, or to the Crimea, as inquiries on these subjects have already been instituted?

We have much more information on the sanitary history of the Crimean campaign than we have on any other. It is a complete example—history does not afford its equal—of an army, after a great disaster arising from neglects, having been brought into the highest state of health and efficiency. It is the whole experiment on a colossal scale. In all other examples, the last step has been wanting to complete the solution of the problem.

We had, in the first seven months of the Crimean campaign, a mortality among the troops at the rate of 60 per cent. per annum from disease alone,—a rate of mortality which exceeds that of the great plague in the population of London, and a higher ratio than the mortality in cholera to the attacks; that is to say, that there died out of the army in the Crimea an annual rate greater than ordinarily die in time of pestilence out of sick.

We had, during the last six months of the war, a mortality among our *sick* not much more than that among our *healthy* Guards at home, and a mortality among our troops in the last five months two-thirds only of what it is among our troops at home.

The mortality among troops of the line at home, when corrected as it ought to be according to the proportion of different ages in the service, has been, on an average of 10 years, 18·7 per 1000 per annum; and among the Guards, 20·4 per 1000 per annum. Comparing this with the Crimean mortality for the last six months of our occupation, we find that the deaths to admissions were 24 per 1000 per annum; and, during five months, viz., January—May, 1856, the mortality among the troops in the Crimea did not exceed 11·5 per 1000 per annum.

Is not this the most complete experiment in army hygiene?

We cannot try this experiment over again for the benefit of inquirers at home, like a chemical experiment. It must be brought forward as a historical example.

Will you now state as nearly as you can how many sick and wounded soldiers were received in the hospitals which you have named at Scutari?

From June, 1854, to June, 1856, about 31,000 are said, according to the Adjutant-General's Returns, made up at the time, 41,000, according to the Director-General's Returns, made up subsequently, to have been admitted at the two hospitals at Scutari above mentioned, to which a third and a smaller one was added, the palace of the Sultan at Haida Pacha, near Scutari, also two ships for convalescents in the Golden Horn at Constantinople, a very small hospital for the artillery stationed at Pera, and a ward over a cavalry stable at Scutari. A certain number of cases also went to the hospitals of Smyrna, Koulali, Abydos, and Renkioi.*

What was the prevailing character of the diseases?

The types of disease, with their causes, may be arranged as follow:—

---

* In 1858, the Director-General sums up the admissions and deaths at the several general hospitals in the East, as follows:—

Name.	Period.	Admissions.	Deaths.
Scutari . . .	June, 1854 to June, 1856	41,325	4,923
Koulali[1] . .	February, 1855 — June, 1855	1,963	509
Varna . . .	June, 1854 — January, 1855	2,846	374
Balaclava . .	October, 1854 — June, 1856	5,686	438
Castle . . .	March 1855 — June, 1856	2,554	96
Camp . . .	April, 1855 — April, 1856	1,083	204
Monastery . .	July, 1855 — June, 1856	911	28
Abydos . . .	December, 1854 — September, 1855	814	82
Smyrna . .	February, 1855 — November, 1856	1,887	154
Renkioi . .	October, 1855 — June, 1856	1,330	50

See Note F., after page 88.

[1] The five months from February to June, 1855, are the only months for which the returns of Koulali are distinguished from those of Scutari.

TIME.	CAUSES.		DISEASES.
1854. November December 1855. January Half February	Exposure Bad food Deficient clothing Fatigue Damp	Scorbutic type.	Diarrhœa. Dysentery. Rheumatism. Frost-bite* Scurvy.
Half February March April	Bad drainage „ ventilation Overcrowding Nuisances Organic effluvia Malaria Damp	Malarial type.	Fever. „ Typhoid. „ Continued. „ Remittent. Diarrhœa. Dysentery. Cholera.

Did it appear to you that many diseases were induced, and others were aggravated in those hospitals?

Typhus attacked both sick and well in hospital. During the month of November, 1854, alone, we had 80 recorded cases of hospital gangrene. Out of 44 secondary amputations of the lower extremities, 36 have died. Cholera broke out frequently both among sick and well. Also, there were frequent relapses of fever and diarrhœa, and the wounded suffered much from both, and, having come in for wounds, they frequently died from fever or diarrhœa.

How many British soldiers died in those hospitals during the period of your residence?

About 4600.

Can you state the number of deaths, month by month, in those hospitals; and the maximum of deaths on any one day?

The following is the head roll of burials for Scutari alone, not including civilians:—

November, 1854	368	October	46
December	667	†November	25
January, 1855	1473	December	7
February	1151	January, 1856	4
March	418	February	4
April	169	March	3
May	76	April	2
June	49	May	1
July	63	June	0
August	60	July	1
September	47		
			4634

* The cases called frost-bite were, according to the sufferers' own account, contracted under circumstances showing no sufficient reason, in the degree of cold, for the effect. They should rather have been called gangrene, from a scorbutic tendency, aggravated by cold.

† This does not include the mortality from cholera in the month of November, 1855, in the German Legion and Osmanli Horse Artillery (which were then quartered in the Barrack Hospital).

The greatest number of burials on any one day was on January 25, viz., 71. On January 26, it was 70. The burials took place, as a rule, within twenty-four hours of the deaths.

You had the best opportunities for observing the condition of the soldier when he entered the hospitals; while he resided in them; when he died and was sent to the cemeteries; when he was sent home an invalid; and when he rejoined the army?

Yes; I was never out of the hospitals.

Will you describe briefly his condition under those various circumstances?

(I.) The sick, from the landing in the Crimea, September, 1854, to the end of January, 1855, were brought to Scutari in 56 transports, some of which were sailing vessels and others steamers.

The sick were received on board, and sometimes detained for several days before the vessel sailed. The voyage for sailing ships was usually six days, and for steamships under two days, but the average time the sick were detained on board was eight days and a half from the time of their embarkation at Balaclava till they were disembarked at Scutari. The numbers which died on board are variously stated, but they are probably in every instance understated.

It appears that about 13,093 sick were embarked during the four months and a half, of whom about 976 died. This is a mortality of no less than 74½ per 1000 in the short space of eight days and a half. Had the embarkation of sick gone on for 12 months, and the same high rate of mortality prevailed among them, the proportion of deaths would have amounted to 3182 per 1000. In other words, the population of the sick transports would have perished on the Black Sea upwards of three times.

The following table gives the monthly details of this catastrophe.

ARRIVED AT SCUTARI.

Months.	Sick embarked arriving at Scutari during the month.	Died on Voyage.	Deaths per 1000 embarked.	Deaths per 1000 on cases treated at Scutari.
1854.				
September 15–30	3,987	311	78	
October	567	15	26·4	
November	2,981	162	54·3	
December	2,656	226	85	
1855.				
January	2,902	262	90	321*
	13,093	976	74·5	198†
February	2,178	41	19	427
3 weeks ending March 17	1,067	5	4·7	315
Ending April 7	860	4	4·6	144

* Viz., from Jan. 7 to Jan. 31, 1855. † Oct. 1, 1854—Jan. 31, 1855.

On a voyage no longer than from Tynemouth to London there died 74 out of every 1000 sick embarked during four months and a half. The loss in individual transports was much greater than this average. The "Caduceus" embarked 430 sick September 23, 1854, and disembarked the survivors at Scutari September 29. During the six days the sick were on board there died no fewer than 114 men, being 26½ per cent. of the total number embarked, or in the ratio of 16.160 per 1000 per annum.

Either these six transports were in a state utterly unfit for sick, as regards overcrowding, ventilation, cleanliness, and attendance; or sick were shipped in a condition utterly unfit for such a voyage. Of cholera cases, those who lived to arrive at all often arrived in a state of collapse only to die.

The superficial space allowed in these transports was fixed by a board of inspection, formed at Balaclava, December 12, 1854, at 6 feet by 2½ feet for sick, and 6 feet by 3 feet for wounded men. The height between deck varied from 6 to 8 feet. One vessel only had a height of 9 feet in some parts, and one a height of 8½ feet.

(II.) During February, it will be seen that the mortality among the cases from the Crimea declined. The per-centage of mortality upon those embarked on board the transports fell from 90 to 19 per 1000. But the mortality at Scutari rose, viz., from 321 to 427. And, even during the three weeks ending March 17, when the mortality on board the transports, which had been, from September 15, 1854, to January 31, 1855, above 74 per 1000, fell to under 5, the mortality at Scutari still maintained its fatal preponderance, and continued at 315 per 1000, half as much again as it had been for the four months, October 1, 1854, to January 31, 1855, which was 198.

(III.) The burial-ground at Scutari was overcharged, and sanitary precautions were not sufficiently observed.

(IV.) The adjutant's table below gives the numbers of men sent home for the months of

October, 1854	90
November	302
December	559
January, 1855	507
February	489
March	1120

It will probably be inferred from this that there were, at first, no sufficient arrangements for relieving the hospitals of invalids, who consequently took up accommodation required for sick; invalids not likely to return to duty are those for whom transport is required. (Upon the sick, removal may have a doubtful result.) It is a well-known fact that, had it not been for sources of private supply, these invalids would have been badly off for clothing, going home.

(V.) The destitution of the men rejoining their regiments, of whom so many had lost their kits, would have been great, had it not also been for sources of private supply. It is well known that they arrived in hospital during the first part of the winter with 'little or no clothing.' I have seen men brought in with nothing on but an old pair of regimental trousers, not their own, a dirty blanket, and a forage cap. In December, 1854, a quartermaster-general's store, for supplying the men going out of hospital with clothing, was established. But it was still found necessary to supply the men with some articles from private sources, upon requisition from the medical officers.

To what do you mainly ascribe the mortality in the hospitals?

To sanitary defects.

Will you now state what the rate of mortality was? And how you have calculated it?

I have calculated it, (1.) on the cases treated; (2.) on the sick population of the hospitals.

I beg to put in a Table.

Average of Weekly States of Sick and Wounded, from October 1st to January 31st, deduced from those given by R. W. Lawson, Deputy Inspector-General, Principal Medical Officer, Scutari.

Date.	No. of Days.	Sick Population of the Hospitals (mean of weekly numbers remaining)	Cases treated (mean of admissions and discharges including Deaths.)	Deaths.	Mortality.	
					Annual rate per cent. per annum on the sick population	Per cent. on cases treated.
1854.						
Oct. 1st to Oct. 14th . .	14	1,993	590	113	148	19·2
Oct. 15th to Nov. 11th . .	28	2,229	2,043	173	101	8·5
Nov. 12th to Dec. 9th . .	28	3,258	1,944	301	121	15·5
Dec. 10th to Jan. 6th (1855)	28	3,701	3,194	572	202	17·9
1855.						
Jan. 7th to Jan. 31st . .	25	4,520	3,072	986	319	32·1
	123	3,140	10,843	2,145	203	19·8

February, 1855.*

Hospitals.	Sick Population.	Cases treated.	Deaths.	Rate of Mortality per cent.	
				Annually on sick Population.	On cases treated.
Scutari and Koulali	4,178	3,112	1,329	415	42·7
Koulali alone . .	648	581	302	608	52·0

* This mortality is considerably understated. If to the Adjutant's Head Roll of Burials for Scutari, page 26, which is probably the correct return, had been added the Medical Officer's Return of Deaths for Koulali, here given, which is the only one we have, the total deaths would have been, not 1329, but 1453; and the rate of deaths on cases treated, not 42·7, but 46·7 per cent.

Statistics of the Barrack, General, Palace, and Koulali Hospitals on the Bosphorus for periods of 21 days, commencing February 24th and ending June 30th, 1855.

Time.	Sick Population of the Hospitals (mean of Numbers remaining at beginning and end).	Cases treated.	Deaths.	Mortality.	
				Annual Rate per cent. per Annum on the Sick Population.	Per cent. on Cases treated.
1855.					
February 25 (and the 20 following days)	3,779	1,621	510	235	31·5
March 18	3,306	1,650	237	125	14·4
April 8	2,803	1,190	127	79	10·7
April 29	2,018	1,350	70	60	5·2
May 20	1,504	996	48	56	4·8
June 10	1,442	1,266	28	34	2·2
126 days . .	2,501	8,070	1,020	118	12·6

I must, however, state that the statistical returns made at various times are very discordant.

Then I find that some of the methods of calculation which I have seen employed give an imperfect idea of the mortality to persons who have not closely studied these matters.

I give an illustration:—

I take a hospital of 1000 beds, constantly occupied during one year, into which, during the year, 10,000 sick are admitted, and 10,000 discharged—viz., 1260 dead, and the rest cured. Then, if we divide the deaths in the year by the *average strength*, as is ordinarily done in determining the mortality of the whole army, it will be found that the 1260 deaths in a year out of a mean strength of 10,000 imply that the mortality was at the rate of 12·6 per *cent.* per *annum.* If this method be employed, it will show the comparative sanitary condition of the sick in two or more such hospitals.

Again, as the number of deaths (1260) took place among 10,000 cases, the mortality was at the rate of 12·6 deaths in every 100 cases.

In the case supposed, one of the imperfect methods of calculation in use, to which I have referred above, if consistently applied, would make the cases 62,000, out of which the 1260 deaths occurred, and the mortality of these cases would be stated at 2 per cent.

The way in which this would happen is sufficiently simple. 1000 cases would remain in the hospital at the beginning of every week; the method objected to counts the cases remaining every week, or fifty-two times in a year, and adds to them the cases admitted, so that, in addition to the 10,000 cases really treated, it obtains 52,000 fictitious cases, simply by counting the same man rather more than six times; that is, by counting the patients remaining 52 times annually, and adding them to the new patients admitted.

If the number remaining at the beginning of every month were counted, 12,000 would be added to the 10,000 real cases treated.

It is evident that, by counting the cases *remaining* at the beginning of every day, or oftener, the *cases* may be multiplied to any conceivable extent.

Again, by dividing the *weekly* deaths by the force in the Crimea, as was done in the published returns, the high mortality in the Crimea *appeared* to be exceedingly low. For people naturally took the mortality per cent. per *week* for the mortality per cent. per *annum*.

For the purposes of accurate comparison it is necessary to reduce all our facts (1.) to unity of time, (2) to unity of strength, or numbers under observation.

I find, after consulting the best statistical authorities, that the mortality of hospitals can be compared in two ways, which mutually check and confirm each other: (1) by dividing the deaths by the mean 'strength' of the sick in hospital, and reducing the mortality to that which would obtain if the time of observation were a year. By this method, it will be seen in the table I give in, that the mortality in the Bosphorus hospitals fell from 41·5 per cent. per annum in February to 23·5 per cent. per annum in the three weeks ending March 17, 1858, to 12·5, 7·9, 6·0, 5·6, and 3·4, in six successive periods of *three weeks* each.

By the second method the *deaths* are divided by the number of *cases treated*. Where the number of sick is stationary, the numbers (1.) admitted and (2.) discharged (including deaths) must be equal in a given time. The number of cases treated will be represented by the numbers either admitted or discharged, as they are equal. When the number of sick increases or decreases, the numbers admitted and discharged will differ; but the number of *cases treated* will, in all ordinary cases, be nearly represented by taking the mean of the numbers admitted and discharged. Thus 6751 patients were *admitted* into the Bosphorus hospitals in the 126 days: February 25—June 30, 1855, and in the same period 9392 patients were discharged. The cases treated are represented by—

$$\begin{array}{r} 6{,}751 \\ 9{,}395 \\ \hline 2)16{,}143 \\ \hline 8{,}072 \end{array}$$

As the deaths were 1020, the mortality on the cases treated was 12·6 per cent.

By this method the mortality of the cases was at the rate of 31·5 in the first three weeks, and fell progressively to

$$\text{Per cent. of cases treated} \quad . \quad . \quad . \quad . \quad \begin{cases} 14{\cdot}4 \\ 10{\cdot}7 \\ 5{\cdot}2 \\ 4{\cdot}8 \\ 2{\cdot}2 \end{cases}$$

As 2501 were under treatment 126 days, the days of sickness were $2501 \times 126 = 315{,}126$ days, which allows—

$$\frac{315{,}126}{8{,}073} = 39 \text{ days}$$

to the treatment of each case on an average.

In the first period the cases were under treatment about 49 days; in the last, 24 days.

While the mortality fell from 31·5 to 2·2, the duration of the cases fell from 49 to 24 days.

Note. (Showing the reason why the cases treated are represented by the mean of the numbers (1.) admitted and (2.) discharged in a given time.)

The number of cases treated ($=$ C) is the number of cases admitted and followed from their commencement to their discharge from the hospital, dead or alive.

When the hospital is (1.) opened and (2.) closed empty, the number of cases treated is the same as the number admitted.

But the hospitals of the Bosphorus contained 3975 sick soldiers on February 25, 1855; 6751 sick soldiers were admitted in 126 days, and on the 30th June 1334 remained under treatment.

Now some of the 3975 cases in the hospital at first ($=$ F) had just entered; others had been one, two, three days under treatment, and others were just on the point of leaving. On an average it may be assumed that they had been treated during half the average term of treatment, and it may be assumed that 3975 treated for a half term —as respects the mortality—may be held to represent half that number, or 1987 cases treated for the full term. To obtain the cases treated, we therefore add 1987 to 6751, the new cases admitted, thus making 8738, from which half the number of 1334 cases remaining on June 30, or 667, must be deducted, leaving 8071 to represent the number of cases treated. Half the 1334 cases remaining ($=667$) ($=\frac{1}{2}$) are deducted, because those 1334 cases have been already counted in the *admissions* as whole cases, although they have only been half their time under treatment, and are fully represented by 667.

The equation will stand thus, putting A for new cases admitted, F for cases remaining on the first day, and L for cases remaining on the last day:—

$$A + \frac{F}{2} - \frac{L}{2} = C.$$

But, by putting D for cases discharged, it may be shown by similar reasoning that

$$D - \frac{F}{2} + \frac{L}{2} = C.$$

And, on adding these two equations together, we have

$$A + D = 2\,C \therefore C = \frac{A + D}{2}$$

The cases treated are nearly represented by the mean between the new cases admitted (A) and the cases discharged (D) dead or alive. It is here assumed that the mortality is at the same rate in the latter half as it is in the first half of its sickness. If it be greater or less, the result will be slightly affected, according as the numbers remaining at the beginning and end of the period of observation are in excess.

From your experience of the great mortality of the hospitals at Scutari, and the manner of estimating it, do you consider the weekly state of the army likely to give a correct idea of the disease and mortality?

From the extracts given in Note A, page 74, which are taken from the letters of the principal medical officer to the commander of the forces, it will at once be seen, that what the commander of the forces and the war department want to know and ought to know about the health of an army in the field was not supplied them.

What they want to know is (1.) how long the army will last at the then rate of mortality.

(2.) Whether the diseases are preventable from which the mortality arises.

(3.) What proportion of the army is inefficient from sickness.

1. The mere information that 'the admissions to strength have been in the ratio of 3·93 per cent.' during 'the present *week*,' and 'the deaths to strength 0·38 per cent,' (*see Note A, page* 75), is simply misleading to the authorities, unless indeed, which is hardly likely, they are thoroughly *au fait* at statistical inquiries.

The standard of comparison all over the civilized world is the per-centage of deaths per *annum;* also the per-centage of admissions into hospital in the same time.

And, without reference to the same standard, it is impossible for the most experienced man to judge.

0·52, 3·93 per cent. look nothing.

But multiply 3·9 by 52 = 2028, in order to get the annual admissions per 1000; and it will be found that the whole force will go twice through hospital in a year, at that rate. And multiply 14·3 (*see Note A, page* 74) by 52 = 7436 per 1000 per annum; and the whole force will go seven times through hospital in a year.

It is not, however, said whether this 'sickness' of '14·31 per cent.' is admissions, or remaining and admissions, or sick population. Probably it is the second. But is this a state of things to congratulate a commander of the forces upon, as being 'pleasing' to him 'to learn?"

Multiply 0·52 by 52 in the same way, and it will be found that the mortality is 270 per 1000 per annum. In other words, that more than one fourth of the whole population of the army will perish in a year.

2. At a time when every one in the Crimea was expecting cholera, which actually did come, and is shortly after recorded by the inspector-general himself, the commander of the forces is congratulated on the 'steadily improving' state of the 'health of the army.'

During the 10 weeks intervening between May 5 and July 14, 1855, 96 per cent. of all the deaths from disease were of the classes usually considered mitigable and zymotic. That is to say that, granting that the four per cent. were not preventible, there might have been saved to the commander of the forces a large part of the 96 out of every hundred men he lost from disease.

Is not this important to him to know from his returns at a glance?

3. For a commander in the field, the number of deaths among his men is, after all, not so necessary to him to know as the rate of inefficient men from sickness he has, and how far the sickness can be mitigated or prevented.

In January, 1855, more than half the infantry were sick in hospital, and three fourths of all that sickness was zymotic, and so it continued from October, 1854, till July, 1855. Nay, even in that dreadful week which included the attack on the Redan, June 18th, 1855, even in that week nearly 65 per cent. of all the deaths in hospital were zymotic.

Of what vital importance it is to a commander and a government to know this?

Now take the unit of per 1000 of the force per annum, and let the advantage of a comparison be shown, viz., of the mortality at different periods of our history in the Crimea.

Deaths per 1000 per annum at different periods.		
January	1855	1173½
	1856	21½
May	1855	203
	1856	8
January 1 to May 31, 1855		628
January 1 to May 31, 1856		11½

Crimea, May, 1856.	Line at home.	Guards at home.
8	18·7	20·4

Thus we were losing in the Crimea, in May, 1856, less than half of what we lose in the line at home, and little more than two-fifths of what we lose in the Guards at home.

It is obvious that what is wanted is for the commander of the forces in the field and the Secretary of State for War to be made instantly and continuously acquainted, (both as to an army at home and in the field) with not only (1) the real proportion of mortality, (2) the real proportion of disease; but also (3) the *kind* of disease and mortality.

Therefore—

1. Present strength.
2. Sick in hospital at a given date.
3. Total admissions since last report.
4. Total deaths since last report.
5. Per-centage of sick to present force.
6. Per-centage of deaths to present force *per annum.*
7. Per-centage of admissions from zymotic disease to total admissions.
8. Per-centage of zymotic cases in hospital to total sick.
9. Per-centage of deaths from zymotic disease, to total deaths from disease.
10. Admissions from wounds and deaths from wounds.

At short periods, both departments should be furnished with a statement of these 10 points, so as to appreciate them at a glance.

It may easily be ascertained what the weekly states for an army in the field are now, as furnished to the commander of the forces by his principal medical officer, and what the daily states furnished by his adjutant-general.

Something of the following I have seen:—

Weekly State.

1. Strength in field as given by adjutant-general this morning.
2. Admissions.
3. Deaths.
4. Present sick.
5. Admissions to strength.
6. Deaths to strength.
7. Sick to strength.
8. Decrease of mortality this week.

But nothing as to 'per annum,' or as to zymotic disease.

Now, is it surprising that the mortality of the army was high in the Crimea? Is it not a fact that the mortality of the army in peace exceeds that of men of corresponding ages in the general population? At home, the Guards and infantry of the line suffer more from consumption and chest diseases than men in civil life, at the same ages, from all diseases put together.

The Guards suffer twice as much from fever—more from fever in some barracks than in others, and more from epidemics in epidemic years than civil life.

There is no *à priori* reason for this. The mere statement of it would excite inquiry into its causes and remedies, if it were generally made known.

This may be summed up in a very few words with the aid of the following data supplied by Dr. Farr:—

No. 1. RATE OF MORTALITY per 1000 per annum.

Effective men of the army at home.	Total . . .	17·5
	Guards . .	20·4
	Line . . .	18·7
Army ages. England and Wales	Country population,	7·7
Army ages.	Whole population,	9·2
Manchester, one of the unhealthiest cities		12·4

No. 2. RELATIVE MORTALITY of the Army at home, and of the English Male Population at corresponding Ages:—

Ages.		Deaths annually to 1000 living.
20—25	Englishmen	8·4
	English soldiers	17·0
25—30	Englishmen	9·2
	English soldiers	18·3
30—35	Englishmen	10·2
	English soldiers	18·4
35—40	Englishmen	11·6
	English soldiers	19·2

That is to say, that if the army were as healthy as the population from which they are drawn, they would die at one-half the rate they die at now.

The army are picked lives, and the inferior lives are thus thrown back among the mass of the population.

The health of the army is continually kept up by an influx of fresh lives, while those which have been used up in the service are also thrown back into the general population, and give a very high mortality.

The general population includes, besides those thus rejected by the army itself (whether in recruiting or invaliding) vagrants, paupers, intemperate persons, the dregs of the race, over whose habits we have little or no control.

The food, clothing, lodging, employment, and nearly all that concerns the sanitary state of the soldier are absolutely under our control, and may be regulated to the minutest particular.

Yet, with all this, the mortality of the army, from which the injured lives are subtracted, is double that of the whole population, to which the injured lives are added.

But, with regard to armies in the field, where half an army may melt away from zymotics alone, or be rendered inefficient just when its services are most wanted, not by wounds, but by disease, of how far more vital importance is it, then, for the authorities to be furnished with such information?

The proportion of admissions and deaths from

Scurvy,
Cholera,
Dysentery,
Diarrhœa, and
Fevers of each class, should be distinctly stated.

For an example of the importance of this, take the following, calculated from official returns for January, 1855.

			Per 1000 per annum to Strength.
Total admissions . .	11,290	hence	4,176
„ Deaths	3,168		1,173·6

	Admissions.	Deaths.
Scurvy	542	31
Scorbutic dysentery .	181	44
	723	75

The deaths from scorbutic disease in that month are put down at 75; a month, when, as is well known, two-thirds of all the deaths and admissions in the army were due to the scorbutic type of disease; for, see the following:—

	Admissions.	Deaths.
Acute dysentery . .	865 }	210 }
Chronic „ . .	*143 }	578 }
Scorbutic „ . .	181	44
Diarrhœa	4,191	1,199
Acute rheumatism . .	342	58
Chronic „ . .	84	9
Frostbite	1,413	124
Scorbutus	542	31
	7,761	2,253

The larger part, if not all, of these 2,253 deaths may be ascribed to the scorbutic type of disease. For 'scorbutic' read 'bad food,' &c., and you have the cause.

---

* The per-centage of mortality on acute and chronic dysentery was perhaps greater than ever has been known in any disease except the worst form of epidemic plague. It was no less than 78 per cent. of the cases. The reason why the deaths from chronic dysentery exceed the admissions appears to be, that many cases of acute dysentery passed into the chronic state before death.

Yet this is not brought forward at all at the time, nor prominently at any subsequent time in our returns.

Again with an army in the field, to give the admission and deaths for the field hospitals is to give no just idea of the mortality, because it is not stated how many of those admissions were sent to and died in the general hospitals at the base of operations.

In the case of the Crimea, till the spring of 1855, no account was rendered of these.

E. g. Published Return.
Crimea.
October 1, 1854—April 1, 1855.

Average strength.	Admissions.	Deaths.
28,623.	52,548.	5,359.

General Return.
Crimea and Scutari.
October 1, 1854—April 30, 1855.

Average strength.	Admissions.*	Deaths.
28,939.	56,057.	10,053.

This mortality is much higher in the early periods than it was in the later periods—higher than it is in civil hospitals, and much higher, too, than the mortality of cases treated in the homes of artisans. Will you state in what respects the sanitary conditions of the soldiers before and after March 17, 1855, *differed from the sanitary conditions of patients in civil hospitals, and of the sick whom you have no doubt visited in their own humble homes?*

The sanitary conditions of the hospitals of Scutari were inferior in point of crowding, ventilation, drainage, and cleanliness, up to the middle of March, 1855, to any civil hospital, or to the poorest homes in the worst parts of the civil population of any large town that I have seen.

After the sanitary works undertaken at that date were executed, I

---

* The admissions at Scutari were not given, because they were nearly all re-admissions from the Crimea.

When we see the loss *thus* given, the real extent of it strikes us. It stands thus:—

	Strength.	Deaths.
Infantry	23,775	9,015
Cavalry	1,915	280
Artillery and Engineers	3,249	568
Undistinguishable . .		190
	28,939	10,053

Sir Alexander Tulloch gives the mortality for those 7 months, at a per-centage for the 7 months. I give it now per annum:—

Strength of the army	28,939
Deaths in 7 months	10,053
Mortality per 1,000 per annum	600

know no buildings in the world which I could compare with them in these points, the original defect of construction of course excepted.

What sanitary defects did you observe in the hospitals at Scutari, in regard to over-crowding, ventilation, cleanliness, and drainage?

(1.) With regard to over-crowding both for barracks and hospitals, the regulation in our service is over-crowding. But at Scutari from the battle of Inkerman till the middle of February, even the regulation space was not adhered to. The men were laid on paillasses on the floor as close as they could lie; there were two rows of beds in the Barrack hospital corridors, where two persons could hardly pass abreast between foot and foot. I have seen 72 patients for a short time in a ward calculated by the inspector-general himself to hold 30.

For full six weeks we had in the Barrack hospital 2000 patients in the space allotted by regulation for 1220; and for six weeks more, 2200 in the space allotted for 1600.

The regulation space is no more than one-half of the space required, if the hospital were in a good sanitary state, but, as the regulation space was actually reduced by one half, we have the following condition of matters:—

1. The space for each patient one-fourth of what it ought to have been.

2. The sanitary condition of the buildings extremely defective.

Can we wonder at the high rate of mortality?

(2.) With regard to ventilation, the hospitals were so insufficiently heated that the inspector-general was of opinion that the men would catch cold by opening the windows. Scarcely any other means of ventilation existed or were provided. It is impossible to describe the state of the atmosphere in the Barrack hospital at night. I have been well acquainted with the dwellings of the worst parts of most of the great cities in Europe, but have never been in any atmosphere which I could compare with it. I have produced statistics to the commission, showing that the mortality on the sick population in January was at the rate of 319 per cent. per annum, the mortality in February at the rate of 415 per cent. per annum, while that on board the transports had fallen from 90 per 1000 in January to 19 in February, showing the improved character of the cases. There is only cause for wonder that the mortality in Scutari was not more, as such an atmosphere as there prevailed was, perhaps, worse than that known to produce cholera and typhus epidemics among healthy persons.

Both ventilation and cubic space were better managed in the General than in the Barrack hospital, but the other causes of disease were the same in both.

(3). The ground about the Palace hospital was very wet, from insufficient draining.

With regard to sewerage, the privies poured their poisons into the corridors and wards; the sewers being loaded with filth, untrapped, and without ventilating openings, the tubes of the privies were their only ventilators, and the sewer gases blew into the wards and corridors. A change of wind so as to blow up the open mouths of the drains, was therefore not unfrequently marked by outbreaks of fever among the patients. And the state of the privies of the Barrack hospital for several months, more than an inch deep in filth, is too horrible to describe.

With regard to cleanliness, the ground about the hospitals, with the exception of the General hospital, was extremely dirty. I have counted as many as six dead dogs just under one ward window. And the Barrack hospital yard, of great extent, was so uneven, muddy, and filthy, as to cause a most unwholesome atmosphere.

The drinking water was not free from organic matter; and on one occasion the dirty hospital dresses have been seen in the tank which supplied water.

A dead horse also lay for some weeks in the aqueduct.

The state of the floors was such, from want of repair, that the medical officer had to choose between leaving them dirty and washing them, which produced erysipelas.

The wards washed in the morning have been seen quite wet at the evening visit.

The walls and ceilings were saturated with organic matter, and should have been lime-washed for safety at least once in three weeks, notwithstanding the inconvenience of such a proceeding for cases which could hardly be moved. The rats, vermin of all kinds, accumulations of dirt and foul air, which harboured under the wooden divans on which the men were laid, rendered the atmosphere still more dangerous to them.

There was a great absence of utensils, and tubs were used for several months in some of the wards, the effect of which on the atmosphere in which the men were need hardly be described.

The burial of the dead was conducted in such a manner, and so close to the General hospital as to affect the purity of the air.

I would refer to our statistics, which show that the mortality rose from 17·9 per cent. in December, on cases treated, to 32·1 per cent. in January, 42·7 per cent. in February; although in February we could scarcely be said to be much over-crowded, and the character of the cases from the Crimea had improved. What else could this increase be due to but to our defective sanitary condition?

How were the sick provided with bedsteads, bedding, ward furniture, utensils, and clothing?

The bedsteads in the General hospital were sufficient, some of theirs having been left by the Turks.

In the Barrack hospital, only a small proportion of the patients had bedsteads in the month of November. There never was a sufficiency of iron bedsteads. Boards and tressels were gradually introduced, till all the patients were raised from the ground.

We had always a sufficiency of blankets, although often covered with filth and vermin. We never had anything but paillasses, excepting for the worst cases, to whom some hair mattresses were given out. The sheets were of canvas, so coarse that the men begged to be left in their blankets. It was indeed impossible to put men in such a state of emaciation into those sheets.

Generally speaking, I have no hesitation in saying that, with the exhausted constitutions of our men, the consequences were fatal of laying them on paillasses, only separated from the floor by the wooden divans in the wards, and by mats in the corridors, which were flagged with limestone, Maltese stone, or unglazed tile. At the same time, I wish to say that none of our men ever made a complaint to me of the hardness of their beds, but the reverse; and I derive my opinion solely from a long habit of watching the sick. Never should a paillasse be seen in a military hospital again without a mattress over it. Any number of mattresses could have been had in an Eastern capital like Constantinople, any day, where every man carries about his bed with him. These would of course have been cotton.

With regard to ward furniture and utensils, there was in the Barrack hospital an absolute deficiency of these things. A soldier is expected by our regulations almost to furnish his own hospital, with some slight assistance from the barrack department. Now, at Scutari, we had neither barrack department nor soldier's kit, for the men had to abandon their kits, as is well known, and did not recover them till a later period. The consequence was, that when we, on two successive occasions, opened newly-repaired divisions of the Barrack hospital to an influx of some hundreds of patients, there was no furniture whatever in the wards but the beds. And all utensils, whether for eating and drinking, taking medicine, cleaning, washing, or other purposes, were deficient. There were also neither chairs, tables, benches, nor any other lamp or candlestick but a bottle. In January, I mentioned this to the then principal medical officer, and he answered, 'I perceive you are not aware that these things are found by the barrack department.' I also mentioned it in the presence of the inspector-general to the purveyor-general, and he answered, that he did not intend to supply the things, having no warrant to that effect, which was true.

I will put in a list of what was in store on January 1, 1855, in the way of ward furniture.

January 1,* 1855.

Plates (tin)	none	Tin drinking cups . . .	none
Candlesticks	none	Urinals (metal)	plenty

* On this day, the number of patients who had arrived in the hospitals of the Bosphorus during the last fortnight amounted to 2532, and were followed by 1044 in the next six days.

Bed-pans	some	Slippers	none
Close stools	plenty [but frames wanted	Knives and forks . . .	none
		Spoons	none
Tin pails for tea	none [at present	Flannel shirts	none
		Socks, pairs	none
Bolsters	none	Drawers, pairs	none
Night-caps	a few	Some tea-pots and coffee-pots.	

With regard to clothing, a limited quantity of very good hospital dresses was in use among the patients. But many of the men came down almost without shirts, or indeed any kind of clothing, but a blanket and a pair of trowsers. From November till after the winter clothing was served out, the supply was very deficient, in hospital, either of shirts, socks, or other articles of underclothing.

The want of shirts, and indeed of almost all clothing, on the part of the men, being proved beyond the possibility of dispute by the statement of almost every official,—in General Airey's words 'the army was almost without clothes,'—What was the 'enormous supply' which, as Commissioner Maxwell states, 'co-existed with constant complaints of the want of the articles in question?'

The number of shirts issued from the quartermaster-general's store at Scutari was, up to 15 February, 4387, a number equal to the men discharged from hospital between December 5, the period when these issues commenced, and January 31, viz., 4349. (*See Note D, page* 83.)

The purveyor states that he issued from October 1 to February 15, 11,234 shirts, of which he says 2500 were delivered to Miss Nightingale. Now, 800 only were delivered to her, the 2500 having been made requisition for by her, but 800 only received. As usual, the amount of the requisition and not of the actual issue was debited. Another example of the fallacy of requisitions as vouchers for receipt, although there is no shadow of suspicion of dishonesty. (*See Note D, page* 83.

The 800 appear again in Miss Nightingale's schedule; therefore deducting from 11,234
2500

8734 remain

Miss Nightingale issued from November 10, 1854, up to February 15, 1855—(*See Cumming-Maxwell Report*, p. 35)

10,537 cotton shirts.
6823 flannel shirts.

17,360

Of which 400 and 400 respectively were, as just stated, from purveyor's store.

Therefore, there were issued—

Shirts.

4,387 from quartermaster-general,
8,734 from purveyor,
17,360 from Miss Nightingale.

30,481

Up to December 4, it appears that the number found in patients' knapsacks was 22! (*See Note D, page* 82.)

Surely, in the above issues there can be found nothing extravagant or unnecessary, but, on the contrary, they must be thought absolutely indispensable, being, as they were, issued upon requisition from the medical officer; the quartermaster-general's issues only excepted, which were made upon inspection of the men's kits.

How were the sick provided with medical comforts, diets, and extra diets?

I do not consider the supply of these things to have been ever deficient in quantity. In quality the bread and meat were good, the eggs, fowls, and milk generally very bad; the arrowroot and port wine sometimes bad. At the same time the cooking was what was chiefly to be complained of. The waste in the wards was enormous, because the men were really unable to eat diets so cooked.

Were the kitchens and extra-diet kitchens sufficient?

There were no extra-diet kitchens at all till those of the female nurses were constructed, which were followed by the construction of one by the excellents taff-surgeon of one of the divisions of the Barrack hospital. The general kitchens simply consisted of boilers and tin pots on Turkish braziers till M. Soyer improved them in April, 1855. Practically, although not theoretically, it was acknowledged by the medical officers, that in order to make the men eat at all, it was necessary to feed them almost exclusively on extras. It need hardly be stated that the extra-diet system is cumbrous, wasteful, inefficient, expensive. (*See Note C, page* 80.)

Were any difficulties experienced in obtaining food, clothing, bedding, medical comforts, or extra diets, when procurable?

Difficulties were experienced in obtaining some articles of extra diet, shirts, clean linen, and bedding, ward furniture and utensils, light and fuel, even when procurable in the market or actually existing in store. I could mention many such examples. After May, 1855, however, no such difficulties, or very few, occurred at Scutari; but in that summer, in the Crimea, and even at a later period, they still existed.

What appeared to be the cause of such difficulties?

I would mention:—

(1.) The system of requisition.

(2.) That of requiring the soldier to bring the contents of his own kit into hospital, which, in war time especially, the time when most required, must oftenest be wanting.

(3.) The fear of being called a 'troublesome fellow,' which, to my positive knowledge, deterred medical officers from making repeated requisitions for articles which they knew to be necessary for their men, or for repairs, because they feared that such conduct would injure their prospects.

This will be denied. But it is true for all that.

(4.) And chiefly, the total want which existed and does exist of any scheme of organization for general hospitals.

Our system has always been regimental; and the only theory of a general hospital which exists in our service is that of an aggregate of regimental hospitals. This will not be denied, because the theory was and is still maintained by many senior medical officers.

How were they overcome, and what, in your opinion, would be the best way of preventing them in future?

(1.) The sanitary defects were first overcome by the Sanitary Commission, which arrived only March 6, 1855.

(2.) With regard to stores, I can best answer the question by putting in an abstract of some of the principal articles supplied from private sources to the hospitals, &c., at Scutari, on requisition from medical officers, as well as to those in the Crimea, and only after ascertaining, in most instances, that the articles did not exist in the purveyor's store or were not to be issued thence.

Article	Unit	Quantity
* Shirts (flannel and cotton)		50,000
Pairs of socks and stockings		23,743
* Pairs of drawers		6,843
* Towels		5,826
* Handkerchiefs		10,044
* Comforters		9,638
Flannel	yards	1,384
Pairs of slippers		3,626
Knives and forks		856
Spoons		2,630
Night caps		4,524
* Gloves and mits	pairs	4,545
Drinking cups		5,477
Tin plates		2,086
* Basins, zinc, &c.		624
Dressing gowns		1,004
Air beds and pillows		232
Thread and tape	packages	74
* Lanterns, candle lamps, and lamps		168
Preserved meats	cases	253
Meat biscuit	barrels	2
Isinglass and gelatine	lbs.	148½
India rubber sheeting	pieces, 325 yds.	26
Camp kitchens, cooking stoves, and canteens		55
Boilers and stewpans		68

Tables and forms,	Brooms and scrubbers,
Baths,	Bedpans,
Soap,	* Tin pails,
Games,	* Combs, scissors. &c. &c.

were supplied with and without requisition.†

The issues of clothing were principally made before the issues of warm clothing took place in the Crimea; and even when these latter issues were being made, a letter dated January 27, 1855, was addressed by the Director-General to the Deputy Secretary-at-War, recommending a clothing store to be formed at Scutari, for men discharged from hospital, which had, however, been done December 5, 1854.

It is but fair to add, for those who do not know the regulations of the service, that the articles marked with an asterisk form no part of hospital furnishing according to the regulations, and that, therefore, the fault was not in those who did not provide them. For the purveyor purveys according to his 'warrants,' but the soldier wants according to his circumstances. The absurdity lies in attempting to provide for war, an abnormal state of society, by normal rules, non-expansive.

(3.) With regard to system, I do not consider that the difficulties were ever overcome at all; all that was done was done by a violent expenditure, and the relaxation of all rule.

No deficiency existed in the purveyor's stores after May, 1855, at Scutari; although I cannot say the same for some of the general hospitals of the Crimea. But this arose not so much from the non-existence as from the non-supply of the stores.

A far more serious question, however, than the want of stores which, with the Anglo-Saxon race, will always be supplied, in such cases, by private interposition, is, the non-organization of a system of general hospital government. For the clash of departments which now constitutes that system cannot be called a hospital government at all. It is a division with grounded arms. 1. The general officer commanding; 2, the quartermaster-general; 3, the adjutant-general; 4, the engineer's department; 5, the paymaster; 6, the commissariat; 7, the contractor; 8, the purveyor; 9, the medical department: these nine departments all step in and appeal to one another to do what each can to make a general hospital march upon regimental contrivances.

Will you state your opinion as to the best manner of organizing a general hospital, so as to make such hospital thoroughly efficient?

---

† A manufacture of stump-pillows, bed-sore pillows—to the number of many thousands,—of arm-slings, and such like surgical needlework, was always being carried on by the nurses. But as this is properly women's work, and it is hoped will always in future be done by them in general, and especially in war hospitals, no further mention of it is here made.

I will reply to this question under the three following heads. In general terms, the organization needed is,

(1.) One for sanitary purposes.
(2.) One for government.
(3.) One for supply.

1. Sanitary Purposes.

The first is really the most important question. The sick might have been loaded with medical comforts, attended by the first medical men of the age; under such sanitary conditions as existed at Scutari, they had not a chance. And yet it does not appear that any adequate means had been applied for remedying the defects, after Dr. Burrell's unfortunate retirement from the service. Nay, what is more, at the time most easy, rather the only time to prevent these evils—viz., in October, 1854, the hospitals in the East were reported 'satisfactory.' This shows the low standard of excellence with which our medical officers were then satisfied; afterwards it was too late. The evils were remedied when the mischief was done. How much easier it would have been to prevent them!

I may quote, in corroboration of this statement, one made by the principal medical officer of the army in the East himself, in the published despatches.

'*Dr. Hall to Lord Raglan, March* 2, 1855.

'* * * *. Out of 442 patients treated in the General hospital at Balaclava, between February 18–24, only three casualties have occurred, which I think may fairly be termed a low rate of mortality, considering the class of patients that are generally sent in there for treatment, men brought down from camp, and found too ill to embark when they arrive at Balaclava.'

The above mortality is 12 per month, or 27 per 1000 per month. If these men had been sent to Scutari in February, there would have died not 12 but 189. The deaths to cases treated at Scutari in February being 427 per 1000, instead of 27 per 1000, and from February 25 to March 17, 315 per 1000.

And, according to Dr. Hall's statement, these cases treated at Balaclava were kept there, because they were in a worse state than those sent down to Scutari.

The distinction between personal and public hygiene must not be overlooked. It is acknowledged in civil life. The officer of health of towns, who was first proposed by Mr. Martin, does not do the work of the physician or surgeon; and, on the other hand, you do not send for your physician or surgeon to drain your street. Had an officer of health been attached to the quartermaster-general's department at the base of operations, who would have put to rights those buildings before we occupied them, how much life would have been spared! This officer should have full powers, through the quartermaster-general's department, to see 1, to the draining of sites; 2, to the sewerage; 3, to the cleansing of outskirts; 4, to the ventilation; 5, to the water supply; 6, to the lime-washing and cleanliness of the buildings; 7, to the allotment of cubic space; 8, to the sanitary conduct of burials.

There is quite enough for one man to do here. There is no more need for this administration of the hygiene of buildings to clash with that of the personal hygiene of the patients, which must be left to the medical officers, than in precisely the analogous case of civil life, where no man ever thinks of confounding the two.

Before and after the works executed by the sanitary commissioners, begun in March, 1855, the hospitals at Scutari bore a similar sanitary difference to that which the gaols of the last century, which were pest-houses, did to Colonel Jebb's prisons of 1857, the most healthy buildings in existence.

The rate of mortality in the hospitals was higher in February than in January, although the mortality on board the sick transports had fallen off from 90 per 1000 in January to 19 in February, proving that the sanitary condition of the hospitals was deteriorating (although the number of sick had diminished), while the sanitary state of the transports and the condition of the sick was improving. The mortality, even to March 17, continued higher in these hospitals than it had been during the average of the four months, October, 1854, November, December, and January, 1855. The sanitary works were then begun, and, after their completion, the mortality fell to less than one-sixth of what it was when the Barrack and General hospitals were first occupied together in October, 1854, and to one-nineteenth of what it was in February, 1855. What other inference can be drawn from such statistics and conditions but this, that, had nothing been done, before four months were out, or as soon as the hot weather set in to give fresh strength to the seeds of epidemic disease buried among us, the hospital population of the Bosphorus would have been swept from the face of the earth, and there would have been no one left in those lazar-houses to tell the tale? Civil life is full of similar lessons.

I am bound to say that the military hospitals I have seen in England, Portsmouth, Chatham, Brompton, are almost as much in want of certain sanitary works as Scutari.

2. Government.

One executive responsible head, it seems to me, is what is wanted in a general hospital, call him governor, commandant, or what you will; and let it be his sole command. Let him be military, medical, or civilian officer, as the possession of administrative talents point out a man for the work.

The departments should not be many:—

1. A governor, solely responsible for everything but medical treatment.

2. A principal medical officer and his staff, relieved of all administrative, and strictly confined to his professional duties.

3. A steward, who should fulfil the duties of purveyor, commissary, and barrackmaster, and supply everything, subject to the governor.

4. A treasurer, who should be banker and paymaster.

5. A captain of hospital attendants, who should undertake the direction of the cooking, washing, care of hospital furniture and government of orderlies.

All these officers to be appointed at home by the War Department.

According to this plan, the governor would cumulate the functions of quartermaster-general and adjutant-general, and therefore the sanitary officer mentioned before would be attached to him instead of to the quartermaster-general. The governor would be solely responsible for carrying out the works advised, and for engaging the labour requisite to do so.

3. Supply.

With regard to the mode of supply, let the steward furnish the hospital according to a fixed scale, previously agreed upon. The mode of supply by requisitions is faulty both ways, both in pretending to supply that which is not in store, and in not supplying that which is. For the requisition remains, although the supply has never been given. And the supply is often not given, although it is in store.

With regard to food, let the steward make contracts, subject to the governor's approval, and with power to buy in the market at the contractor's expense, if the contractor fails.

A scheme of diets should be constructed, according to the most approved authorities, in order to save the cumbrous machinery of extra diet rolls.

Equivalents might be laid down, so as to afford the necessary choice, depending on the nature of the climate, the season of the year, the state of the market, the productions of the country, &c.

To what extent and in what manner could female nursing be rendered available in general hospitals attached to an army in the field or at home?

Female nursing might be introduced in general hospitals both at home and in the field, if only women of the efficiency, responsibility, and character of head nurses in civil hospitals be appointed. Say one to not less than 25 bad cases; the orderlies doing under the head female nurse the duty done in civil hospitals by assistant-nurses.

But the head female nurse must be in charge of all that pertains to the bed-side of the patient, of his cleanliness, that of his bed and utensils, of the administration of medicine, of food, of the minor dressings not performed by the surgeon, in short, of all that concerns the personal obedience of the patient to the orders of the surgeon. She must accompany the surgeon on his visits, and receive his orders. She must also be in charge of the ventilation and warming of the ward. She must report any disobedience of the orderlies, as far as regards the patient's personal treatment.

There need be no clashing with the ward-master or hospital serjeant. On the contrary, it would be the duty of these to enforce the nurse's authority. They will have enough to do, besides, with returns and accounts, and with enforcing discipline as to hours, meals, clothing, &c., among the orderlies out of ward.

The female nurses should be of course under a female head, whose duties must be carefully arranged so as to be in accordance with the code of hospital regulations.

Are you of opinion that female nursing could be employed advantageously in regimental hospitals?

I am not.

What was the method in use at Scutari, when you went there, for receiving and admitting sick and registering their admissions and diseases, and how were the results, deaths, and burials recorded?

Up to the middle of February, 1855, the method said to exist was the following:—The adjutant or his serjeant stood at the landing-place or hospital gate and took the names of those who were able to speak, receiving a verbal report from the officer in charge of the number who had died during the voyage. At first, there was no arrangement for registering the sick when they were received into hospital. And the erasure of the man's name from the diet roll was at one time the only record of his death, excepting the adjutant's head roll of burials.

I believe that the medical officer in charge always endeavoured to make a nominal roll on the voyage from Balaklava to Scutari; more than one has stated that it took him half a day from his professional duties about the sick to do so. But, whatever this nominal roll was, it went to the principal medical officer's office, and not into the wards; one copy was given, I have been told, to the deputy assistant quartermaster-general on landing. But at what period this was first done I am not aware. It has been stated to me by several medical officers that, even as late as March, 1855, the inspecting medical officer of transports, upon going on board, frequently found this list imperfect, owing to the illness of the medical officer in charge, or other unavoidable cause, and had to complete it then and there.

Till the middle of February, or later, we had no bed-tickets, and no regular death returns. It is well known that the sick were put on board at Balaclava without nominal or numerical lists up to February, 1855. Upon their admission at Scutari, an assistant-surgeon has been charged late in the evening to take the names, regiment, regimental number, &c., and to make the diagnosis of above 100 patients before nine the next morning. One-twentieth have been known to die before they could give their names.

After the end of March all this confusion ceased. But it was not till the end of April that a census of the hospitals was taken, and returns "squared" with the adjutant-general in the Crimea.

Returns possibly existed of which I knew nothing. I am speaking only, first, of what I saw; second, of what I was informed of by witnesses on the spot and at the time.

The conclusion I draw from all this is, not that any one was in fault, for all were burdened with work beyond their strength, but that there was at first no system of registration for general hospitals, and that a registrar is an essentially necessary officer for such an institution.

In consequence of the great confusion of the hospitals from the battle of the Alma till a late period of the winter of 1854–55, a number of men were buried from the hospitals, exceeding in six months the deaths reported in the orderly room by 530. But the deaths reported in the orderly room exceed those reported in the hospital books,

In November by	12
„ December „	143
„ January „	125
	280

while, as if to make up the lost ground, the medical returns, which had exceeded the orderly room returns

by 12 deaths in October, exceed again
by 253 „ February,

265

bringing the sum of deaths very nearly equal. The adjutant having buried 280 men more than the total number reported by the inspector-general as dead, the account had to be balanced at a subsequent period.

By the census taken of the hospitals, April 30, 1855, it appeared that 517 non-commissioned officers and soldiers had been buried whose deaths were unrecorded. An official report was then sent up to the adjutant-general in the Crimea, and they were struck off the strength of the army. The names and regiments of 28 others remained unknown.

I have carefully compared the statistics from six different official sources, and none of them agree.

It is possible that, in some of these cases, the numbers of deaths may be the numbers occurring in the several months, in other cases the numbers recorded in those months. At the same time, the great discrepancies in the several numbers shake confidence in their accuracy, and render it difficult to make any use of them for statistical purposes.

From details given in Note B, page 77, it will be seen (1.) that there are three different returns of the deaths on board the transports from Balaclava to Scutari, and that the number reported is different in each separate return. Nay, there is scarcely a single ship for which the returns agree.

2. The adjutant's head-roll of burials, the most trustworthy record of deaths, exceeded in six months by upwards of 500 the number

reported as having died in the hospitals. And the adjutant states that this list of burials does not include civilians. (See Note B, page 78.)

3. The director-general himself represents the very numbers in the 'return of total sick *treated in Lord Raglan's army*' as having been '*treated at Scutari*' alone in November, December, 1854, and January, 1855. (See Note, page 78).

4. The returns of dead at Scutari from October 1 to December 31, 1854, vary as follows:—

Adjutant's return of burials.	Reported deaths, Orderly Room.	Principal Medical Officer's return.
1301	1046	910

Smyrna hospital did not then exist. And Koulali did not exist till December, and contained then, as a maximum, but 240 patients.

These sources of discrepancy are therefore eliminated.

The result of my examination of these statistics is simply this, that however satisfactory they may be to the departments who have put them forth, and whichever of them may be correct, exhibiting as they do such palpable diversity, still, to any one not in the secret as to how things which apparently differ so widely are nevertheless identically the same, they convey no trustworthy idea as to the sickness and mortality of the army in the East, and that, for any practical purpose, they are, as put forth to the public who are most interested in the matter, not absolute truths, but only approximations.

The calculations which I give unquestionably understate the mortality actually experienced.*

How were discipline and obedience to medical and hospital orders enforced among the patients and orderlies of the hospitals in the East?

In the sense in which discipline is understood in the army, *i.e.*, of teaching or learning how intelligently to obey orders (which we saw

---

* Our total loss in the Crimean War (of all arms), from first to last, was—

Men killed	2,598	according to adjutant-general.
,, died in hospital	18,058	,, director-general.
	20,656	
Officers killed (157), died (233) .	390	,, adjutant-general.

But in making up this total of 18,058, died in hospital, the Scutari mortality given by Dr. Andrew Smith for the six months, October, 1854, to March, 1855, exceeds by nearly 500 the mortality reported to him by his own medical officers at the time, as cited at pages 29, 30. And Dr. Andrew Smith gives the numbers who died on board the transports for the five months, October, 1854, to February, 1855, as exceeding by 177 the numbers reported officially by the surgeons on board at the time, see page 27. So that I have understated the Scutari and transport mortality by at least a number of nearly 700 men; something like a whole regiment, of which we may say that we shall never know the how, the when, nor the where of its perishing.

carried to the highest perfection in the Crimea), there was little or none in the hospitals.

It is not meant by this that the men made a 'shindy' whenever they were left alone, or that they did not stand at attention when the medical officers went into the ward. There was, in fact, a little too much of this. It is meant that hospital discipline consists in enforcing hospital treatment, and hospital treatment consists in the administration of medicine, of diet, the application of the minor dressings, the cleanliness of the ward, that of the patient, his bedding, and his utensils, his personal obedience to the surgeon's orders as to keeping his ward, his bed, or his position, &c. &c. To enforce in all these matters obedience to the surgeon's orders is to enforce hospital discipline.

Another branch of discipline consists, no doubt, in the man's respectful demeanour to his officers, in the accuracy of diet rolls, returns, accounts, &c. &c. This is not here spoken of, because it is not strictly hospital work. In hospital work, the meaning of the words to be 'in charge' is, we presume, to receive the medical officer's orders, to see them obeyed, and to take measures so that they cannot be disobeyed without its being known and remedied.

Now, in military hospitals, there is no one *thus* 'in charge.'

The ward-master or hospital-serjeant has duties so multifarious that three men could not perform them all with any satisfaction, and the nursing work is generally sacrificed to the writing work, unless the ward-master be, as sometimes, but rarely, happens, a very good nurse.

The orderlies do not bring 'skilled labour' to the work, and the medical staff corps far less. There is little or no training; it is a truism that less work is done less well in the same time by unskilled than by skilled labour.

The cleaning and airing of the wards in the morning would make a housemaid laugh; at Scutari each orderly worked at it in his own way, and then the patients undid it all, and it had all to be done over again.*

Except when the medicines were given by the medical officers

---

* The above remarks by no means apply to the war-hospitals alone. They apply often as much to many military hospitals at home and in peace. It is true that the 'outside of the platter' is sometimes licked tolerably clean for inspection. But has the visitor ever pulled out the dirty linen crammed under the patient's mattress? or ever found under the patient's pillow his tobacco, his pipe, sometimes lighted, his clasp-knife, dirty shirt, spoon, and towel, if he has one? These are pulled out by the medical officer, who is obliged to confiscate them for the time being, and desire the hospital serjeant 'to take them away,' well knowing that neither he nor the patient has anywhere else to put them.

Cleanliness may be attained by extraordinary exertions for an occasion in the military hospital as it is now; but order could only be obtained by a miracle, since hardly one of the appurtenances which are necessary for order is ever to be found in a military hospital. The institution of the ward-scullery is one almost entirely unknown; and there are appliances of civilization considered too advanced for the themselves or by the women, they were taken by the patient or not, at his own discretion.

The meals were at first irregular, and throughout the patients scrambled for them. It occurred frequently that the bad cases, when unable to feed themselves, were not fed at all, except by the women; that a great mess of cold arrowroot and wine stood by the bedside the whole day, till it was thrown away; that the poultices were put on cold, or left on till they were hard, and then not washed off; that the bed-sores were unattended to till they had become so bad that the medical officers' attention was called to them; that the patients were left dirty unless they asked to be washed; that the utensils were never emptied more than once a day; that the keeping certain very bad cases constantly dry and clean was almost wholly unattended to; that the patients ordered to remain in bed were as often out of bed, or even out of the ward, or smoking in the privies when these were cleansed and ventilated. The position of surgical cases in bed was generally not attended to.

Many orderly medical officers were so zealous in their duties that they would drudge through all these details, and see them carried out themselves; but this is not discipline, nor can it be, unless the medical officer were in his ward through the whole 24 hours.

If these things are denied, I say I speak from personal experience, and, with regard also to the military hospitals at home, I have seen and heard of the same things there.

It would be easy for me, in all these assertions, to cite the particular instances, but I am not disposed to bring upon individuals censure for defects which I believe to be inherent in the system.

The medical staff corps are, for the most part, nothing but sweeps and ration-carriers. There are some first-rate subjects among them, but the endowments of most of them were first-rate ignorance of the details of nursing.

If all the nurses were turned out of civil hospitals and men engaged promiscuously, without character, to do their work, we should see the same results.

How were the inspections conducted, as far as they concerned hospital order?

The medical inspections, as far as I saw them, with some very brilliant exceptions, regarded only hospital order, and were conducted with little reference to the state of the patients.

I will describe the state of the General hospital at Balaclava when I came into it, April 11th, 1856, a time when there was by no means any pressure.

---

soldier, such as the ward-press, lockers, properly fitted-up lavatories, &c. &c. It appears to be supposed that the men will be more orderly and cleanly abroad, where they cannot have all these appliances, for having been subjected at home to the want of them; the English soldier is always desirous to be clean where he can.

The first night I found from ten to twenty men of an Irish regimen talking and drinking in the extra-diet kitchen.

A quantity of extra bedding and clothes were cumbering the wards, instead of being in store.

Under the beds and under the mattresses were the patients' private clothes, large boxes, carpenters' tools, wood, coals, dusters, brooms, stones used in cleaning the wards, boots, shoes, and slippers. These things were concealed by the bedclothes.

One patient, suffering from frost-bite, who subsequently died, had not been moved for a week. Being unable to leave his bed, and having been neglected, he was found in a state indescribably horrible.

The excellent second-class staff surgeon, who came on duty the same day I did, immediately set to work to introduce real order and cleanliness. On April 14, the principal medical officer of the army visited the hospital without requiring this staff surgeon, who was in charge, to accompany him; and on the subsequent day the principal medical officer censured this officer for 'the state of dirt and disorder,' which was the removal of 'dirt and disorder,' in which he had 'found the whole establishment.'

In that hospital, when going over at night the whole of the wards, of which seven were in the building, and fourteen were huts, I have not found a single orderly perfectly sober who was sitting up in the huts, and one or two in the building not sober either. This was the medical staff corps.

Can you state why the homes of the poor in the country are kept comparatively clean and healthy, on very moderate means?

I think that the woman is superior in skill to the man in all points of sanitary domestic economy, and more particularly in cleanliness and tidiness. I think great sanitary civil reformers will always tell us that they look to the woman to carry out practically their hygienic reforms. She has a superior aptitude in *nursing* the well quite as much as in nursing the sick. At the same time, I am bound to say that nothing can be more perfect, at least to outward appearance, than the cleanliness of a ship. But the sailor is a race *à part.*

Is it the peculiar skill and industry of the English labourer's wife to which this is referable in the one case, and to the incompetency of men on the other to conduct the domestic economy of a home or an hospital?

I think so. I think the Anglo-Saxon would be very sorry to turn women out of his own house, or out of civil hospitals, hotels, institutions of all kinds, and substitute men-housekeepers and men-matrons. The contrast between even naval hospitals, where there are female nurses, and military hospitals, where there are none, is most striking in point of order and cleanliness.*

---

* I should perhaps state that there is a great difference, generally speaking, among the women of Great Britain and Ireland in this respect.

I would put the Anglo-Saxon race in the southern and north-western counties

Are there not matrons in all the best civil hospitals?

In all that I am acquainted with.

II.—Hospital Construction.

Will you state to the Commission what you consider to be the best plan of hospital construction, for fulfilling the requisites of good sanitary condition and facility of administration, with your reasons for preferring any plan or plans to others?

The best principle of hospital construction is that of separate pavilions, placed side by side, or in line. The former is preferable. It diminishes the distance to traverse from block to block.

The distance between the blocks should be not less than double the height.

There should not be more than two flats to the block, nor more than one ward to each flat.

There is, however, no objection to having 70 to 80 sick under one roof, if, for the sake of economy, it be necessary to build each pavilion with three flats instead of two, although two flats are more convenient for administration.

For the purposes of administration, the building ought to be in a square; the basement story connected all round by an arched corridor, with an open terrace above.

The whole hospital should be erected upon an arched basement.

A hospital formed of separate pavilions could be built in line, provided large, roomy, well-ventilated, and well-lighted staircases intervened between each two pavilions.

This is the plan of the new military hospital at Vincennes, which, however, forms three sides of a square.

This principle would have been illustrated in Netley, had it been made so thin as only to have admitted two rows of beds between the windows, in the breadth of the ward, had there been no corridors, and had the pavilions been separated by staircases. It would then have been perfectly healthy; but administration would have been very difficult in so long a line.

As it is, there are to be two complete and different administrations, with two kitchens, two sets of offices complete; it is the most expensive and least *administrable* form.

There is only one kitchen to Vincennes.

That not more that 100 patients can with safety and facility of administration be massed under one roof, has come to be an acknowledged principle of hospital construction.

Buildings of two flats are most compatible with perfect sanitary

---

first in point of domestic management; far below these come the Danish race in the eastern counties and the mixed race in the manufacturing counties, and last, the Irish and Highland Celt.

conditions and facility of administration. Even in emigrant ships the occupation is limited now to one deck.*

Will you state what you consider the number of sick which a ward should contain, for health, discipline, and administration, with your reasons for preferring one number to another?

The best size of wards for ensuring the two conditions of health and facility of discipline, is from 20 to 32 sick.

Wards smaller than of 20 beds multiply both the attendance, unnecessarily, and the corners, unfavourably for ventilation, in proportion to the number of patients. Wards larger than of 32 beds are undesirable, because more difficult and expensive to ventilate.

Wards, again, smaller than of 20 beds are more difficult to ventilate by natural means alone. A certain amount of space is requisite for diffusion, in order to secure perfect natural ventilation.

The mode of construction in hospitals is, it is presumed, to be determined by that which is best for the recovery of the sick. If any other consideration is taken, such or such a per-centage of mortality is to be sacrificed to that other consideration.

But it so happens that the safest for the sick is the most economical mode of construction.

Take the example of Portsmouth hospital, or Netley, where the windows are at each end of the ward. There should not properly be more than four beds in each of those wards. For it is undesirable ever to allow more than two beds between each two windows; otherwise, when the windows are opened, the effluvia blow over all the intervening row of beds before escaping.

Wards of a small size are also decidedly objectionable, because unfavourable to discipline, inasmuch as a small number of men, when placed together in the same ward, more readily associate together for any breach of discipline than a larger number.

It has been proved by experience that the presence of head nurses, whether male or female, one to each ward, is essential to discipline, and a sufficient number of such nurses cannot be allotted in smaller wards. One head nurse can easily overlook 40 patients in one large ward. In four small ones it is almost impossible.

In the event of a death taking place in the ward, the survivors, when they are few in number, are far more likely to be affected by it than a larger number.

---

* The utmost simplicity of plan is an essential of good hospital construction. Complication of plan interferes with light, ventilation, discipline, facility of supervision. Every hole and corner, every passage, every small ward, which need not have been there, interferes with these four vital conditions of a good hospital. Every skulking place which can be spared must be avoided. As an invariable hospital rule, rather more than elsewhere in military hospitals, publicity may be considered as the best police and the best protection. It is far better that 30 patients should see the nurse's door than one or none. It is quite necessary that the chief ward attendant should be able to see the whole of his patients at once.

It is very desirable, for the purposes of discipline, that men of the same regiment should not be placed together in the same small wards of general hospitals.

Discipline.—There needs only a comparison between the discipline of civil and military hospitals to substantiate the above assertions. If discipline mean the enforcing obedience to orders by teaching how they are to be obeyed, there is little or none in a military hospital.

In the administration of food, of medicine, in the cleanliness of the patient's person, bedding, and utensils, in the patient's personal obedience to medical orders, as to rising or remaining in bed—all of these being matters about which there is no question or demur in a civil hospital, the discipline in a military hospital is far inferior.

How can it be otherwise? Unless the medical officer is converted into nurse, cook, and housemaid, he cannot see to all these things. The ward-master or hospital-serjeant is necessarily absent from his ward a great deal, on account of returns and accounts. The orderly cannot enforce obedience. There is no one *in charge* to enforce the medical orders, *i.e.*, without whose knowledge no disobedience can take place, as there is in a civil hospital, where one 'ward sister' is daily seen in absolute control of 30–40 men, as far as obtaining absolute obedience to medical orders goes; there is no rebellion, tacit or open.

Two other considerations are involved in that of the size of wards, economy and clinical instruction.*

It may be asked, why should not all the sick be placed in one ward, provided there be cubic space enough? The answer is, with from 20–32 sick a height of 15–17 feet is enough, but it would not be enough for more, and height always involves expense.

The greatest economy and the greatest safety to patients is in the above number.

Also, without the most perfect ventilation, there is always more danger of effluvia being driven by a draught till it accumulates in one part of a very large ward, as was the case in the long corridors of Scutari.

Wards of a moderate size, like those indicated, are better for the purposes of ventilation than wards half the size; and are less subject

---

* The question, what is the best number of beds in a ward? has been but little considered in England in regard to economy and efficiency of service—hardly at all in military hospitals. The more beds in one ward, the fewer the attendants necessary in proportion, and, within a certain limit, the greater the facility of supervision. But the sanitary necessities of cubic space per bed, &c., impose the limit. You may make your ward too large for the chief attendant to overlook the whole at a glance, which he and still more *she* ought to be able to do. Allotting a sufficiency of superficial area, &c., 32 beds per ward seems to be this limit assigned by European experience. See, for further details on economy of hospital nursing, Note E, page 85.

to a hospital atmosphere than wards of double the size. But a ward of this latter size may be rendered perfectly healthy by having a height in proportion to its width.

Where clinical instruction is intended, to admit even a class of six students into a ward of 12 sick is increasing the population in the cubic space by one-half. There is more than twice the room proportionally for students, in a ward of double the size. On the other hand, if the number of students be very large, a ward of 20 patients, it must be at once admitted, is too small. The ward must be increased, and with it its height and its cubic space; for, be it remembered, the whole of the proportions of the ward, not only its length, must increase with its number of beds; for, if the ward be very long, in proportion to its height and breadth, it becomes not a ward but a corridor, and all corridors are objectionable for sick, because it is impossible to ventilate them safely; because, in admitting air, the effluvia may be driven from one end and be accumulated at the opposite end faster than it can be taken out. The right proportion is a fixed one. But in a ward for 40 patients, 20 bad cases will be disturbed while 20 slight cases are being examined. If 20 sick only be put in each ward, the slight cases may be put together.

What amount of cubic space should be allotted for each bed?

The cubic space for each patient in this climate has been fixed by European sanitary science at not less than 1500 feet.

A good proportion for a ward of 20 patients would be 80 feet long, 25 feet wide, and 16 feet high. This would give 1600 cubic feet to each bed. It would give 13 feet between foot and foot, which is not too much where there is a clinical school. It would give an average of 16 feet to each 2 beds in width.

Half the sick are supposed to be on each side the ward.

What is the best proportion of windows to beds, and what ought to be the relative position of windows and beds, with your reasons for preferring one arrangement to another?

One window should be allotted for every two beds; the window to be not less than 4 feet 8 inches wide, within 2 or 3 feet of the floor, so that the patient can see out, and up to the ceiling.

The pair of beds between the windows to be not less than 3 feet apart. With a very bad fever case, I would leave the other bed empty, for the sake of isolating the patient. Miasma may be said, roughly speaking, to diminish as the square of the distance. With good ventilation, it is not found to extend much beyond 3 feet from the patient; although miasma from the excretions may extend a considerably greater distance.

Windows are to be placed opposite each other, and to be either double or filled with plate glass; the former is preferable, as it affords the opportunity of indirect ventilation in all weather. Wire-gauze across the open part of the window would afford an extent of surface for ventilation not otherwise to be obtained, and preclude all possibility of draught upon the patient.

Windows opening as at Middlesex and Guy's Hospital, in three or more parts, with an iron casting outside, to prevent a delirious patient from throwing himself out, are the best form of plate-glass window.

No part of the ward ought to be dark. This is of the utmost importance, in many cases. The light can always be modified for individual patients. But even for such patients to have light in the ward is not the less important.

There are three reasons for this multiplicity of windows:—

1. Light.
2. Ventilation.
3. To enable patients to read in bed.

The necessity of light is established by all scientific inquiry and experience. The proportion of windows to cubic space is of the first importance to health. It has been lost sight of in English architecture, owing to the unfortunate window-tax, which has left its legacy in giving us a far smaller proportion of light than in French houses. In huts the proportion of window space to cubic space is far greater than in buildings. One main cause of the unhealthiness of large numbers of men congregated in one large building, even with sufficient cubic space, is the disproportionately small window space. In the huts in the Crimea, during the last 22 weeks of our occupation, the mortality of the whole army was only two-thirds of what it is in England.

For the same purpose of ensuring a sufficiency of light, the walls should always be white, excepting perhaps for some few cases of ophthalmia.

What is the best material for the internal walls and ceilings of wards?

Impervious walls are of the first importance for hospitals. These walls should be of Parian or other similar cement, or glazed tiles. Brick, used at Portsmouth hospital, is highly objectionable from its porous character. Plaster is objectionable from the same circumstance, it absorbs organic matter. Both require very frequent lime-washing to keep them healthy.

What is the best material for the flooring of wards, and your reasons for preferring one material to another?

The materials used for floors may be oak wood, pine wood, composition, and tiles.

Oak wood, well seasoned, is the best. No sawdust or other organic matter capable of rotting should be placed underneath the floor. Concrete, or some similar indestructible substance, would be the best for the purpose.

The reason for using oak wood is, that it is capable of absorbing but a very small quantity of water. And it is very desirable to diminish even that capability, by saturating it with beeswax and turpentine. Beeswax is an inalterable substance.*

The joints of the flooring must be fitted well together, and cemented with marine glue, or any other impervious substance. The object is, of course, to prevent any water from entering the floor.

Impervious, non-absorbent cement or composition would make a capital floor, used as it is in Italian houses. But, on account of its great conducting power, it would be necessary to furnish each patient with a pair of list shoes, and a small bedside carpet.

The stairs and landings should be of stone. The corridors should be floored with diamond-shaped flags or tiles, which stand better than those laid in the usual manner. The terrace might be either covered with asphalte or glazed tile.

What accommodation for nurses, extra diets, clothing, and clean linen should be attached to the wards?

There should be a nurse's room, and a small scullery attached to each ward, also a press in the ward.†

It is perhaps hardly necessary to add that there should be a few small wards for 'casualty' cases, for noisy or offensive cases. But they should be completely separated from the other wards, and under a completely appointed staff of their own, not attached one to each larger ward, which renders proper attendance extremely difficult.

What kind of baths should be attached to the hospital?

The baths should be separated from the pavilions, but connected by the corridor.

The walls and ceilings should be of pure white cement, or some similar material, the floors of tile. They should be suitably ventilated and warmed. They should contain hot and cold water baths, sulphureous water, hot air, medicated and vapour baths, shower baths, and douche. There should be a portable bath to each ward.

What is the best form of hospital kitchen?

The kitchen should be placed away from the wards. Its walls and ceiling should be of pure white cement, for plaster has a tendency to fall off, from the vapour and effluvia of the kitchen.

---

* This floor should be cleaned like the French *parquet*, by *frottage*.

A hospital floor should never be scoured. A very good hospital floor is that used at Berlin, which is oiled, lackered and polished so as to resemble French polish. It is wet-rubbed and dry-rubbed every morning, which removes the dust. Its only objection is want of durability.

† A scullery, with complete, efficient, simple apparatus for its various purposes, places for washing up and cleaning, and for *ward* cookery, so that the nurse can warm the drinks, prepare fomentations, &c., without jostling the orderlies who are washing up, is a great promoter of order, efficiency, and work.

It is a doubtful arrangement to have a clothes' room for each ward. A military hospital should have but one clothes' room, under charge of some man. Room for storing and issuing dried clean linen, as well as laundry-room, should be provided. Foul linen should be delivered twice daily into the laundry. A large box in the scullery is the least bad place for it in the meantime.

The cooking apparatus, boilers, &c., if placed in the centre of the kitchen, instead of against the walls, will afford twice the amount of fire space.

In the Paris kitchens there is a brick erection in the middle of the floor, with iron doors and brass mountings, coppers with covers, places for baking and for roasting, &c.

The dressers are against the walls.

The floors are flagged with square flags.

This appears to be the most convenient mode of erection.*

What is the best form of laundry for a hospital?

The excellent new washing, drying, and wringing machines lately invented are so numerous that it would take too long to enumerate them. On the whole, the laundry at the Wellington Barracks, which also washes for all the Guards' hospitals and barracks, and the new laundry at Haslar naval hospital, are the best I have seen. But every day brings in fresh inventions, and a reformer is always adopting the good ones.†

It is a further question in army matters, whether you should train a body of men to do as much as possible by hand, so as to be serviceable in the field, where machines cannot be had, or whether you should make use of all the inventions for saving labour which are now so good, and daily improving. Probably both must be done.‡

---

* For camp hospital kitchens Soyer's boilers and Little's, or Smith's of Glasgow 'enchantress' stoves, are those I prefer from experience. They cook well, and secure both great economy in fuel and great capability of varying the diets.

† I do not think that any reliable comparison has yet been made between the French system adopted at the Salpétrière and Lariboisière hospitals and the English system. The French consists in filtering hot ley through the clothes, which are placed for that purpose in large tubs, with a compartment at the bottom, from which the ley is pumped up by machinery, and allowed to flow over the top of the linen, through which it filters into the compartment, to be again raised by the machine.

This plan is stated to be the most economical which has been tried in Paris.

There are several good plans in use in the British hospitals. The essential characteristic of the Haslar one is boiling by steam, the linen being afterwards placed in a rotating washing machine.

Another method in use at the Wellington Barracks, where the washing of the Guards' barracks and hospitals is done, consists in passing the linen through slowly rotating washing tubs, in which it undergoes a process of *waulking* by wooden rods. This latter plan is both economical and effectual.

To ascertain which is the most really economical of the French and English plans, it would be necessary to inquire not only into the relative cost of washing, but into the relative wear and tear. The hardness or softness of the water must also be taken into account. The softer water is the cheaper, both in the consumption of soap and the wear and tear. Now the Paris water is in hardness to that of London as 20 to 16, and as 20 to 2 to that of Glasgow. Probably the Paris method is the only economical one with the Paris water. But it takes about five times the superficial area of the Guards' method.

‡ One thing, however, cannot be too strongly said. There is scarcely a laundry I know in any military hospital at home where there is any sufficient provision either for washing or for preserving the health of the washers. This latter must be attended to primarily. In almost all military hospitals in England, it is true, the heavy washing is done, nobody knows how, by contract. But the lighter washing

How should nurses and orderlies be accommodated?

If orderlies are to sleep among their patients, the per-centage of mortality will be of course raised among them. This was the case at Scutari, where it was very high, though it will never be known how high. Statistics are, however, not necessary to establish such an obvious fact. The orderlies should sleep at a distance from the wards, or, if sanctioned by military authority, in little rooms adjoining their wards, and they should not take their meals with the patients.*

With regard to the female nurses, there is the further question as to what class of nurses should be employed in military hospitals. My own opinion is entirely against employing any but women of the efficiency, responsibility, and character of head nurses in civil hospitals. For such a nurse a small airy room off her ward, and looking into it, so that she can always have it under command, is the best for her efficiency, and need not be injurious to her health.

Unless, however, there are facile means of access to another nurse's room, in case of illness, there must be only a day room for each head nurse adjacent to her ward. She must sleep at a distance from her ward, and contiguous to the other nurses. Assistant female nurses are better not employed.

What kind of bedstead and bedding is the best?

No bedding but the hair mattress has yet been discovered that is fit for hospitals. Hair is indestructible. It does not readily retain miasma. And, if it does, heat easily disinfects it. It may be washed. It is not hard to the patient. It saves the objectionable use of a blanket UNDER the patient.

Straw paillasses are absolutely objectionable. They are cold; and, in some cases, the abstraction of heat from the spine lowers the patient's vital energy to a degree which does not leave him a chance. I am of opinion that the loss of life must have been great during the war from laying our patients on paillasses, which were either placed on wooden divans, or on the flagged corridors, with only a mat between.

The bedstead should be iron; and there are great differences in the way the sacking is put in. There should always be a shelf at the head. The French have one at the feet too.

The naval and civil hospitals have all kinds of dropsy and surgical bedsteads for raising a patient when he cannot be moved, for inclining him at a certain angle, &c., also water and air beds. There is no

---

is done in some miserable lean-to, without any arrangement for 'getting-up,' drying, or airing the linen, which is done, if at all, at the ward-fire. This is simply destruction to anything which can be called nursing.

* A scullery—small, but not too small—attached to each ward is, as has been said, essential to order, cleanliness, and discipline. It should be well provided with cold, and, if possible, with hot water. Each orderly should have his locker, each his safe in it, with a key of his own, and he should have his meals there, if the military authorities are not against it. No patient should enter the scullery unless sent there to wash up, &c.; and as a rule none should be sent there.

reason but a general objection to comfort to prevent us using these bedsteads in military hospitals.

What ward furniture should you recommend?

Oak furniture decidedly. White window curtains are used in some French hospitals, not to exclude the light, but to look cheerful. They are desirable, but not necessary.

The less ward furniture, speaking generally, the better. But hitherto, as is well known, a military patient has been expected almost to furnish his own ward.*

Do you wish to say anything about water supply or drainage?

The water must either be drawn from a tank at a distance from the hospital, or from a main under pressure; but never from a cistern within the hospital.

The fault of the water supply in Parisian hospitals is, that the water is either carried up by 'porteurs d'eau,' or pumped up and remains standing.

There is no question that this is wrong.

There should be convenient means in or close to the wards for obtaining pure drinking water for the sick.

No drain should ever pass under a hospital; all sinks, water-closets, lavatories, and baths, should be so placed that the drainage should be conveyed directly away, without passing under any part of the hospital.

All drains or pipes for the purpose of conveying away water from any part of the hospital should be carefully trapped between the outer wall of the building and the sewer; and all drains should be ventilated.

What is the best position for the water-closets and lavatories with regard to the wards?

The water-closets should be placed at the end of the ward, opposite the entrance, and separated by a lighted and ventilated lobby. They should be of the best construction, self-acting. Adjoining should be a small bath room for bad cases, and lavatory.†

---

* The material of the different utensils required for ward service should be settled. The use of glass or earthenware for all eating, drinking, and washing vessels is recommended from its great superiority in cleanliness, and in saving time and labour in cleaning. Tin vessels of certain kinds cannot, by any amount of cleaning, be freed from smell.

† The lavatory should have a row of white earthenware basins fixed in a stand, with outlet tubes and plugs, each basin should have a hot and cold water pipe, and there should be not less than one to each six or eight beds. There should also be in the lavatory a hot and cold water pipe, from which the portable bath can be filled.

The sink, which should have a partition of its own, adjoining the water-closet, should be a high, large, deep, round, pierced basin of earthenware, above a *large* hole, with a cock extending far enough over the sink for the stream of water to fall directly into the vessel to be cleansed. This is far preferable to the usual oblong sink. The scullery sink is, of course, to be entirely separate, and for entirely different purposes from this.

What do you consider the best system of ventilation for a hospital, and why?

The doors, windows, and fire-places should be the means of ventilation for such wards as these; nothing else is wanted. If a hospital must be ventilated artificially it betrays a defect of original construction which no artificial ventilation can compensate; it is an expensive and inefficient means of doing that which can be done cheaply and efficiently by constructing your building so as to admit the open air around.

But there are buildings of original defective construction which it is undoubtedly necessary to ventilate artificially. And in countries where fuel is dear and cold severe, the problem complicates itself, because it is a less consumption of fuel to warm the fresh air as you admit it.

In the case supposed, warming is a necessary part of ventilation, and heating the air required heats the wards, without extra cost of fire-places, which always burn many times the quantity of fuel expended for warming.

In England, where fuel is cheap, there can be no such excuse.

If attendants cannot be trained to keep the rooms ventilated without draughts, there is a defect of intelligence, and attendance on the sick is not their calling.

Occasionally, ventilating shafts carried up from the ceiling of the ward to the roof will be found an advantageous means of renewing the air.

There should be one or more open fire-places in the ward, but lofty, so that the throat of the chimney shall be above the patient's head and bed.

We look upon the chimney as the best ventilating shaft; and one disadvantage of artificial ventilation is, that you must then supply the fire with its own consumption of air by a shaft to itself. Or it will take the air from the artificial ventilation and cause it to cease to act. Whereas with natural ventilation the fire sets it acting, takes the air from the room, and is the most valuable means of ventilation.

Our grandfathers' lofty fire-places are the greatest loss in modern house architecture. The little low fire-places of this date bring the best current of air below the stratum in which we are breathing. With our system, to breathe the best air, we must not be more than six years old, or we must lie down.

What do you consider the best system of warming for a hospital, and why?

Radiation; open fire-places. Heated air from metal surfaces should never be used for warming. It has a tendency to produce disease of the lungs. The hot-house system of warming with a brick floor and brick flues is perfectly safe; but an earthenware floor in hospitals, unless glazed, is inadmissible, because of the great absorption of water.

In a ward of the size mentioned, 80 ft. by 25, two fire-places would probably be necessary.

III. Organization of General Hospitals.

Do you advise the establishment of additional general hospitals?

Yes; at Malta, Corfu, Gibraltar, London, Portsmouth, Aldershot, and perhaps at Shorncliffe and Fermoy, there are sufficient bodies of troops for general hospitals to be formed; and it would be most desirable for the interests of science, economy, and administration to establish on stations where larger numbers of troops than one regiment are collected, at home and abroad, permanent divisional or brigade hospitals, together with a permanent administrative staff of officers and subordinates, for the treatment of the sick of the station and of detachments within reasonable distance.

The efficient hospital staff, trained under such an organization, would supply regiments going on foreign service.

The medical staff would consist permanently of an inspector and deputy-inspector, the rest of the staff would not be permanent, because regiments change on the station. The complement of a regiment is one surgeon and two or even three assistant surgeons. One of these would be appointed to serve on the staff of the general hospital, one would remain with the regiment as sanitary officer, and the third might be set free on leave to go to some military general hospital for the purpose of study, for, say that there were five regiments, five surgeons would easily do the duty for the sick in a general hospital, where now, in regimental hospitals, it is the business of fifteen.

Do you prefer general hospitals to agglomerated regimental hospitals, and why?

The system of agglomerated regimental hospitals is no system at all, if it means merely assigning one or more wards in the same building to each regimental hospital, with the same kitchen to all, so that if there are ten hospitals, there may be ten surgeries, there are ten hospital serjeants, ten staffs of medical officers, ten of everything, while in one hospital there may be two severe cases, in another twenty.

The only theory of a general hospital which we heard during the late war was that of making it as much like a congeries of regimental hospitals as possible; and, though this theory neither was nor could be systematically carried out into practice, it remained and does still remain the theory, and prevents, as it then prevented, any attempt to devise a better and more practicable scheme. For,

1. The surgeon cannot be relieved of his multifarious non-professional duties.

2. No comparison of different treatments, with a view to the furtherance of science or the advancement of individual knowledge, can be made, while in the military medical service the organization remains exclusively regimental.

Without actual experience, indeed, it is evident that to divide a small number of sick into half a dozen hospitals, which have all to be kept up as complete and separate establishments, instead of uniting them into one, necessarily involves considerable waste of labour and expense.

It is contrary to every analogy in England. What should we say if every parish in Middlesex insisted upon sending its own lunatics under their own surgeon to be separately lodged in Hanwell Asylum? We should say no improvement in science, classification, or administration is here possible.

In what particulars is improvement impossible under the present exclusive system of regimental hospitals?

1. The management and internal economy of hospitals, cooking, washing, supply and issue of extras, &c.

2. The financial part of the service.

3. The professional and economical superintendence.

4. The organization and superintendence of the attendance department.

5. The supply and dispensing of medicines and medical appliances.

6. The introduction of proper baths, lavatories, &c.

7. The classification and treatment of particular diseases.

8. The imperatively needed organization of hospitals for soldiers' wives and children when sick, the want of which is most demoralizing.

9. The arrangements for the treatment of sick prisoners and others not belonging to regiments on the station.

In large permanent establishments only can these things be properly regulated.

How does this system affect the army medical officers?

The standard of professional service can never be raised among army medical officers under the regimental system.

For to do this, three things are necessary :—

1. Sufficient professional occupation.

2. Command over large numbers of cases of the same affections.

3. Opportunity for comparing results of different modes of treatment.

In what other ways does the existing system obstruct the progress and diffusion of medical science?

Under the present system,

1. No practice can ever be established as the best; the sick being treated in so many hundred hospitals by as many hundred surgeons.

2. No system of mutual information and clinical instruction can be established. Individual experience remains individual property, or is buried in official reports.

3. No professional specialties can be acquired. For there is no opportunity of collecting the necessary material for study, practice, and experiment in special diseases. Nor is the experience of one man shared by others.

4. No publicity can be introduced, bringing individuals and esta-

blishments into competition with their rivals in civil life and other armies.

5. No prizes, distinctions, or promotion for scientific and professional labours can be systematically introduced.

What are the principal objections to the regimental hospital or agglomeration of regimental hospitals?

1. The impossibility of professional progress.
2. The impossibility of any organization of government.
3. The impossibility of any organization of attendance.

The agglomeration of the regimental hospitals does not modify these evils any further than what was actually done in the Crimea, where the hospital huts of each regiment were placed as near as possible.

As to 2. and 3., the management and government of the regimental hospitals are now entirely vested in the medical officer and hospital serjeant, under the military authorities.

Can advantage be taken of the division of labour?

It is easy to see that no division of labour is possible.

The duties of the hospital serjeant, by regulation, are so multifarious that it amounts to a physical impossibility for one man, however active and intelligent, to perform them all.

He is Ward Master;
Serjeant;
Steward;
Head Nurse;
sometimes Dispenser;
sometimes Purveyor's Clerk.

The consequence is that there is no nursing, and no hospital discipline, in the sense in which these things are understood in a civil hospital.

Again, the regimental system was shown to be in the Crimea utterly incapable of meeting the requirements of an army in the field under the circumstances, and yet we might never again be in circumstances in the field so favourable to the formation of regimental hospitals.

When the general hospitals came to be formed, there were but two or three medical officers who had any idea of organizing a system for 1000 to 2500 sick.

What reasons are usually alleged in favour of the regimental hospitals?

1. That the regimental surgeon knows all his men; and
2. That the sick soldier does not like to be separated from his regiment.

Are these reasons valid?

No; only partially. 1. The fact of the frequent changes of the regimental surgeon, especially in time of war, is overlooked. Even promotion takes him away from his regiment. The hospital serjeant is frequently the only permanent officer. He sometimes knows all his men; but often the surgeon does not.

2. As far as I have seen, nothing but evil comes of comrades lying together sick. They are depressed by each others' sufferings or death. They are led to combine against hospital discipline. I incline to think that the task of governing a number of men in a civil hospital, who have never been under any discipline at all, and are there merely as casual inmates, is considerably facilitated by their being strangers to one another.

Is the regimental hospital system a preparation for organizing a general hospital in time of war? And is it safe to depend upon an exclusively regimental system, when, in all emergencies, the general hospitals are what we must depend upon?

The primary necessity of management in institutions containing, say, 1000 persons, is the reduction of departments to their smallest possible number, and the definition of their functions.

This has not even been attempted. The nine departments of our war general hospitals have been described. Before a patient could eat his dinner in the Scutari general hospitals, it had to be manufactured through the medium of the commandant who assigned the orderlies and cook; of the engineers' department who repaired the kitchen; of the purveyor who supplied a portion of the food; of the commissary who, through the contractor, supplied bread, meat, and fuel; and of the soldier himself, who supplied, out of his own kit, some of the utensils for eating and drinking.

A question of hospital repairs has been known to pass from the medical officer to the purveyor, thence to the principal medical officer, back to the purveyor, thence to the quartermaster-general's department, commandant, engineers' department, and was ultimately decided upon without appeal to the informing judgment, viz., the surgeon in charge, who saw his patient suffering from (say) a leaky roof or broken pane of glass, without the means of redress.

It is hardly necessary to state that it is not safe to depend on any system which leads to such results.

What has been your experience of the working of the present system of written requisitions, checks, and counter-checks?

This system of checks and counter-checks seems to have been invented for the purpose of saving money instead of for that of saving the lives of the sick. Now it fails in its object both ways, because the lives of men are of more money value to the country than any saving can ever by any possibility be in such matters; and also because it actually wastes money; for the clerk system and check system require such a staff as to cost far more than the additional supplies would do.

No system can be more expensive than ours, for these reasons.

It is also inconsistent with prompt and efficient action, and consequently hazardous to the sick.

Can you suggest any simpler and more efficient arrangement?

A governor, who should be responsible for everything; a principal medical officer; a steward, who should supply everything, like the agent in the naval hospitals; a captain of hospital attendants, who should overlook the whole attendance department; a treasurer; and a sanitary officer attached to the governor. What more do we want? We must have regimental hospitals also. It is not meant to do away with a necessary, at all events, and, in some respects, an admirable system. But a medical officer must be conversant with both, otherwise, as surely as we again have war, shall we again break down in our general hospitals.

Has experience established the necessity of general hospitals?

Yes.

It does not appear necessary to discuss whether there ought or ought not to be general hospitals, because there must be general hospitals. If they are an evil, let the evil be made as light as possible by good organization; if they are good, let them be made as good as possible.

But, in what war have our general hospitals not been a signal failure? In the early part of George II.'s reign, at Coimbra, and, most fatal example of all, in the late war.

If this has been the case, and if this always will be the case, reasoning from the past, is it not advisable at least to go into the question of constituting a scheme of government for general hospitals?

Certainly.

General hospitals, at the base of operations, when this is at a distance from the army, must necessarily fall through, between the two authorities, the commander of the forces engaged in active operations, and the departments at home, unless there is a governing head with commensurate responsibility on the spot.

If it is said that the general officer commanding in the district is the governing power, it is obvious that the commander of the forces will give the man least wanted and least efficient in the front for this purpose; whereas, if a governor for the hospital itself were appointed from home for his known powers in administration, the man most efficient might be sent.

Ought not the principal medical officer to be the governor?

Not if it can be avoided. It is obvious that the real object and glory of the medical profession are medical skill and practice: and, if a good medical officer be put to do something else, his services are lost in that which he has spent his life to acquire.

But experience is the true test, and if the experience of the failure of the Scutari hospitals will not teach us at least to discuss the question, no argument stronger can be used.

How should the general hospital be governed and organized?

An executive head, who is there on the spot to answer for failure, whose business it is without any question to govern this institution and nothing else, with power corresponding to the responsibility, which is his and no one's else, this is what we seek in all similar known institutions. He ought to have full power to obtain

Labour,
Transport,
Supplies.

Labour he must have power to hire, when not obtainable from the military train, also transport, as far as land is concerned, when not obtainable from the land transport corps; the sea transport he must have from the admiral on the station. Supplies he must have power to obtain without reference to the commissariat, who are not responsible.

Whether his steward is to be a commissariat officer, detached for the purpose, or an officer from a different service, it is not my province to discuss.

But it cannot be disputed that the articles of supply for a hospital are different from those for an army. The French take a different bread for their hospitals. We are allowed to give 1*d.* per pound more for our hospital meat.

It cannot be disputed that it is better to let the responsibility for all supplies rest upon one person than upon two.

What is the difference between a civil hospital and an agglomeration of regimental hospitals?

The former has one treasurer, who administers the finance of all the wards. It has one committee, who manages the supply of all the wards. It has one secretary, who manages the correspondence for all the wards. It has one house governor or superintendent for all the wards; one kitchen which cooks for all the wards; one laundry which washes for all the wards; one surgery; one apothecary; one pharmacy and laboratory for all the wards; one set of attendants, under one head; one set of accounts; one system of supply. It has one set of officers; one operating theatre; one dissecting room.

There is not a bad case in the hospital which may not have the advantage of the advice of all the best officers in the institution; there is not a surgeon or physician who may not be called into consultation by all the rest. I have never known an instance where the interests of a patient were sacrificed to the interests of an individual officer. So strong is the professional spirit in civil medical life that the experience of each man becomes the common property of all the rest.

I have recently heard it stated, for the first time, that each civil surgeon is situated as regards his ward precisely as a regimental surgeon is in his hospital. I am bound to record, as strongly as I can, that the reverse has been my experience in a great variety of instances.

What peculiar advantages would the patient enjoy in general hospitals well organized?

In general hospitals, there can or ought to be no difficulty in obtaining all that is necessary for the patient. In all regimental hospitals, it is impossible for the regimental surgeon to obtain all that he wants for his patients; unless there is such a staff for each regimental hospital of 50 patients, as there is, or ought to be, for a general hospital of 500; which is manifestly impossible.

What is the real objection to general hospitals in the minds of army medical officers?

A most legitimate and unanswerable one under the present system.

A regimental surgeon is master, as he ought to be, of his own practice in his regimental hospital. The practising surgeon in the general hospital practically is not. Both are subject to inspection, but practically the difference is great.

With the slightest knowledge of human nature, with the commonest sense, it is obvious that the inspections, as now conducted in general hospitals, the system of superintendence and subordination of scientific acquirement to rank, must destroy in a man his sense of responsibility, and (the most important quality, the latest acquired, in a man of action, which a medical man must be) his self-reliance.

Therefore do surgeons under the present regime most reasonably and justly prefer being regimental surgeons.

What are the legitimate objects of inspections?

Inspections ought to be merely for the purpose of collecting facts, with the view of enabling the authorities to act upon them.

Your whole purpose in educating and examining the medical officer is to fit him for medical practice. If he is not fit to practise, he is not fit to be a medical officer; and yet you would subject him to weekly interference in his practice and in petty details much better left to himself.

The only rationale of superintendence is, when it comes from a superintendent better qualified than the superintended, which the treating surgeon, in the army medical department, does not feel to be the case in most instances. For the superintendent has, in most instances, ceased to practise.

Are any peculiar views of their proper sphere of action entertained by any section of army medical officers?

There is a portion of the army medical department who like to keep the housekeeping and book-keeping in their own hands. This is a very small portion. But they do not consider how impossible it is to raise the true dignity of the department, to attract the best men into it, as long as this is the case.

To make high-priced labour do what low-priced labour can do, to make men at 7*s.* 6*d.* a day do what men at 1*s.* 6*d.* a day could just as well perform, is the true way to degrade a scientific department.

And I do not see how it is possible, without the organization of large permanent establishments, in which alone a proper administrative staff could be trained, to make the army surgeon other than a sort of maid of all-work.

Are the confidential reports attended with any evil consequences?

The evil inflicted by them is not so much the injury to individuals, severely as this is sometimes felt, as the lowering of the whole moral tone of the department, the danger of depriving a man of his conscientiousness, his sense of responsibility, his true dignity. For army medical men are like all other medical men. And such interference with their practice and their reputation would be fatal to all. No increase of pay will raise the department while these things are suffered to exist.

What is the best class of men to be recruited for the medical staff corps?

Discharged soldiers or civilians of good character. They should always be men who can read and write.

In what manner would you propose to have them instructed in their duties?

If they are to nurse, unquestionably they should be instructed in the duties of nursing, as also in those of cleaning, &c., by head female nurses, who understand these things. To the superior ranks, a few simple anatomical lectures might be given by the medical officers. Dispensers must, of course, be suitably educated, if they are to dispense.

What are the duties of orderlies attending on the sick?

And what would you propose to be the respective duties and relative position of the nurses and orderlies employed in the same wards?

The administration of diets and of medicine, the making of poultices, and application of leeches, blisters, and the minor dressings, the management of the ventilation and warming of the ward, should always be in charge of the head-nurse. Under her, the cleanliness of the ward, bed, bedding, and utensils of the patients, as well as his personal cleanliness; the fetching of diets, the warming and ventilation of the wards, are to be attended to by the orderlies, but always under the nurse's surveillance. And she is to have the power of reporting disobedience on their part.

What should be the proportion of orderlies to sick?

Not less than one to each seven sick, where there is no female head-nurse. I am not speaking of convalescents.*

* Where there is a female head-nurse, but no lifts or supply of hot and cold water all over the building, still one orderly to every seven patients will not be too much. Upon an average, these two appliances make the difference of one orderly's duty to a ward of thirty sick. And a ward of thirty sick, with these appliances, will be better served by three orderlies and half a nurse (for one head-nurse could see after two such wards) than a ward without the appliances would be by four orderlies. The proportion of severe cases in military hospitals being generally much smaller than in civil ones, one nurse could overlook two wards of thirty beds, provided they were on the same floor. But, in all other respects, the wards should be quite separate.

How long would you keep a man on duty in the ward?

Not more than twelve hours altogether. Watches of four hours are good, as in the naval service. But this is a question which the proper military authorities alone can decide.*

How would you provide for the discipline and regulation of the orderlies?

What duties should you assign to the ward-master?

What proportion of ward-masters to the number of orderlies employed?

There should be a ward-master for every five or six wards, whose whole business should be to see to the regulation and discipline of the orderlies, and to the enforcing obedience to the orders of the nurse. But it is obvious that if the nurse is not in authority in regard to all that concerns the patient, her duties will become impossible. If she is in authority the orderlies will willingly obey.

It should, therefore, be compulsory that she should report a refractory orderly to the ward-master or to the captain of orderlies, if such an officer be created.

The ward-master should be in charge of all returns, accounts, states,

---

* Twelve hours for the orderly to be off duty are not at all too much. Upon an average, all men and women, after a laborious day, require a good night, in the long run. When they do not have it, either health, efficiency, or sobriety, or all three go. A strong soldier is no exception to the general rule. In the long run, if made to do night duty after a laborious day, he will either go to sleep, or drink to keep awake, or he will get knocked up before his time. It is sound economy to give watchers sufficient sleep. You will get more, and get it longer, out of the man by giving him twelve hours on duty and twelve hours off. It is better for him to have eight than seven hours' sleep; and one or two hours for exercise and fresh air each afternoon, or each alternate afternoon, make a man last longer than going to and from his bed and his ward—care, of course, being taken that he does not exercise himself in some tap. Supposing regular night duty required in a ward of thirty men, supplied as above, and served by half a nurse and three orderlies, it might be worked thus:—The principal medical officer will decide whether the same orderly should do the night duty for a week, or the three on successive nights; probably the latter. The orderly might come on night duty at 9 P.M. and remain on duty until 9 A.M., thus taking his share in the heavy morning work of cleaning the ward, &c. A large ward, got into thorough order by 9 A.M. is in very good time. A night ration for night watchers is, I believe, indispensable. A nurse, whether male or female, watching and fasting in a ward from nine to nine, or even from nine till six, would either soon be unfit for duty, or put drams in his or her pocket, or doze through the night. It should be arranged that the nurses who sleep before and after the watch can do so quietly. This is by no means always attended to, either in civil or military hospitals. The following extraordinary system of night nursing is that which prevails in the Army at present:—'Nursing is managed by comrade-patients told off in three watches of two hours each for the night.' 'Orderlies are likewise warned, and often sit up for the purpose.' Or they 'arrange among themselves,' without any exemption meanwhile from day duty. Or soldiers are sent in from the ranks to attend specially upon each bad case. Or it has been proposed that the patients shall '*nurse each other!*' All this refers to things at home, not to war-time, nor to any emergency. Upon each and all of these systems, or no systems, it is hardly necessary to make any comment.

diet-rolls, &c., so as to set the nurse completely at liberty to attend to her ward.

What objection is there to the hospital with a long corridor and small wards?

The least administrable form of hospital is the long corridor with wards of from eight to ten patients, opening off one side. Attendance, meaning, of course, due attention to the patients, suitably superintended, becomes almost impossible, especially at night.

How do you propose to obviate the evident danger of collecting a great number of sick men, labouring under diseases of every kind, into one large building where they breathe a common atmosphere?

The French break up their general hospitals into separate blocks, while we sometimes agglomerate our regimental hospitals into one building, thus *reversing* the sanitary law.

The Lariboisière hospital at Paris has 600 patients under six roofs; and we had in the Crimea accommodation for 500 or 600 patients at the Castle hospital in from 20 to 30 large and small huts.

One administration, in both cases, for all.

This is the true principle. All that has to be manufactured, as the cooking, washing, &c., should be, as much as possible, concentrated into one; while human beings, sick or well, should be distributed as much as possible.

In London, one wash-house will do for a number of families, and is as good as giving an additional room to each; but this does not break up our house system and convert London into a gigantic public institution.

NOTE A.

No. 47.—Lord Raglan to Lord Panmure.

(*Received March* 20.)

(Extract.) Before Sebastopol, March 3, 1855.

I beg now to lay before your lordship Dr. Hall's report of the state of the sick, and I will direct him to make one weekly, which I will transmit for your lordship's information.

(Enclosure.)

Dr. Hall to Lord Raglan.

Before Sebastopol, March 19, 1855.

My Lord,—In transmitting the weekly state of sick of the army to the 17th instant, I have the honour to state, that though the sickness still amounts to 14·31 per cent., the mortality does not exceed 0·51 per cent., which is a proof that the diseases are milder in character; and I think I may safely say the general health and appearance of the men is greatly improved; and had not the duty, by the unavoidable operations of the siege going on, been increased of late, I think the sick list would have been still more diminished, as the men's condition is in every other way so much improved, both in diet, dress, and accommodation.

(Enclosure.)

DR. HALL TO LORD RAGLAN.

Before Sebastopol, April 3, 1855.

MY LORD,—In transmitting the weekly state of sick to the 31st March, I have the honour to state, that I am sure it will be pleasing to your lordship to learn, that the general health of the army continues steadily to improve; and although fevers and bowel complaints continue to prevail, they are both assuming a milder character, and the latter are of much less frequent occurrence.

During the present week the admissions to strength have been in the ratio of 3·93 per cent., and the deaths to strength 0·38 per cent.

Last week, the admissions to strength were 4·35 per cent., and the deaths 0·52 per cent.; which makes a decrease of 139 in the admissions, and 43 in the deaths during the week.

(Enclosure.)

DR. HALL TO LORD RAGLAN.

Before Sebastopol, May 28, 1855.

MY LORD,—The enclosed weekly state of sick to the 26th instant, I am glad to say, shows an improvement in the sanitary condition of the army. The cases of cholera which have been admitted during the week have been of a milder character, and the mortality from that disease has been much less; but it has extended to the Sardinian contingent, to the men of the land transport corps, and to the shipping in the harbour of Balaklava, and, from these sources, the admissions and deaths in the general hospital there have been considerably increased; no fewer than eighteen out of the twenty-five casualties which occurred there, having taken place amongst these extra patients. In future, I will have the extra patients excluded from the weekly state and shown separately; but as I cannot get fresh returns made out in time for the post on the present occasion, I have deducted the numbers from the general return.

The admissions to strength during the present week, have been in the ratio of 4·20 per cent.; and the deaths to strength, 0·27 per cent. Last week they were 4·53, and 0·47, respectively.

Fevers have been less numerous during the week, but diarrhœa has been slightly on the increase; and it has been noticed that many convalescent from fever have been seized, in some of whom the disease has run on to cholera and terminated fatally.

I have, &c.
J. Hall,
Inspector-General of Hospitals.

Dr. Hall to General Codrington.

Head Quarters, Camp, Crimea, Dec. 31, 1855.

The following abstract shows the admissions and deaths during the week, and those during the previous week.

	This week.		Last week.	
	Admitted.	Died.	Admitted.	Died.
Fever	182	15	173	16
Head affections	9	2	2	2
Chest ditto	279	0	221	1
Diarrhœa	117	3	204	4
Cholera	3	3	12	7
Dysentery	8	1	40	0
Rheumatism	44	2	58	1
Frost Bites	98	0	269	1
Wounds and Injuries	72	4	48	6
Ophthalmia	37	0	22	0
Other diseases	425	6	374	2
Total	1,274	36	1,423	40

I have, &c.,
J. Hall,
Inspector-General of Hospitals.

NOTE B.

I have carefully compared all the statistics from six different official sources. I will here put in four sets of returns.

I.

Name of Ship.*	Deaths according to		
	Cumming-Maxwell's Return.	House of Commons Return.	Adjutant's Return.
Kangaroo	—	22	—
Dunbar	22	10	30
Cambria	—	21	—
Vulcan	18	10	25
Andes	15	4	—
Colombo	30	30	57
Arthur the Great	24	50	30
Orient	32 or 33	26	45
Caduceus	114	104	—
Courier	16	33	16
Cornwall	6	6	8
Negotiator	6	6	—
Lady McNaughten	3	3	3
Australia	8	3	12
Cambria	None.	—	—
Echunga	7	6	1
Palmerston	—	7	5
&c. &c.			

* The above comparison is made, taking the first seventeen ships in order, as their names stand in the returns.

II.

From Returns of Adjutant's Office at Scutari.

Months.	No. of Burials in each month.	No. of Deaths reported in each month.	Excess of Burials over Deaths in each month.
1854, September	165	78	87
,, October	266	219	47
,, November	368	291	77
,, December	667	536	131
1855, January	1,473	1,360	113
,, February	1,151	1,076	75
,, March	418	416	2
,, April	165	152	13
,, May	76	76	
	4,749	4,204	545

REMARKS.—517 of the 545 have been struck off the strength of the general depôt, the other 28 men were brought ashore either dead or insensible, with no marks to ascertain their names or regiments.

III.

Appendix to Report from the Sebastopol Committee, 2nd Report, p. 705.

Return showing the Total number of Sick and Wounded treated in Hospital at Scutari.

Months.	Total Treated.
During *November* 1854	16,846
„ *December* 1854	19,479
„ *January* 1855	23,076

A. Smith, M.D., Director-General.

13, St. James's-place,
29th March, 1855.

Appendix to Report from the Sebastopol Committee, p. 470.

No. 1.—Return showing the Total Number of Men of Lord Raglan's Army Sick during each Month, from the Landing in Turkey.

Months.	Total Sick or Wounded of all arms during each month.
1854, April	503
„ May	1,835
„ June	3,498
„ July	6,937
„ August	11,236
„ September	11,693
„ October	11,988
„ *November*	16,846
„ *December*	19,479
1855, *January*	23,076
To 17 February . . Crimea . .	9,284
„ 25 „ . . Scutari . .	6,725
„ 17 „ . . Abydos . .	385
„ 25 „ . . Gallipoli .	70
„ 20 „ . . Smyrna . .	500
Total to latest dates in February	16,964

Army and Ordnance
Medical Department,
14th March, 1855.

A. Smith,
Director-General.

IV.

Deaths at Scutari and Koulali Hospitals, October 1, 1854, to April 30, 1855.

	1. Commandant at Scutari.	2. Medical Returns.*	3. Head Roll Burials.	4. Reported Deaths.	5. Depôt Returns.	6. Inspector-General, as quoted by Sir A. Tulloch. (Infantry).
October . .	—	250	266	219	213	144
November .	—	267	368	291	244	228
December .	—	393	667	536	493	423
January . .	1,480	1,235	1,473	1,360	1,079	1,193
February . .	1,254	1,329	1,151	1,076	1,254	1,261
March . . .	424	555	418	416	324	587
April . . .	—	200	165	152	213	216
						4,052
						† Rem. 470
	—	4,229	4,508	4,050	3,820	4,522
	At Scutari and Koulali.		At Scutari.			

† Being Cavalry, Artillery, and Royal Engineers.

The above are the returns from different official sources. The discrepancies in them will no doubt be explained to us at some future time. Smyrna is not included. The Smyrna hospital was opened to sick February 15, and from that day till March 31, 127 deaths occurred, by the inspector-general's return. Returns 1 and 2 purport expressly to be for Scutari and Koulali. Returns 3, 4, 5 to be for Scutari alone. Return 6 *may* include Smyrna. Otherwise, being for the 'infantry' alone, it would prove, in defiance of Euclid, that a part is greater than the whole. For the deaths in the month of March of infantry alone are greater than those recorded by the other returns in March for all arms of the service. However, the Smyrna supposition by no means accounts for the difference—as return 6 is something less than return 2 (the medical return) for February, although something more for March.

---

* For the first four months in Return 2, read—

October 1 to November 4	250
November 5 to December 2	267
December 3 to December 30	393
December 31 to January 31	1,235

NOTE C.

Extra-Diet Kitchen.

The kitchen apparatus of the Barrack and General hospitals, as left by the Turks, consisted of thirteen large copper boilers in the former building, and eleven in the latter, set in brickwork, eight and five of which only, respectively, were available with safety.

In the General hospital the brickwork was loose, and the fuel was supplied from within the building, so that the kitchen was filled with smoke and the cooking impeded. This state of things continued till after the arrival of M. Soyer, in April, 1855, when the engineer officer set the boilers properly, and placed the flues so that they could be lighted from without—at the same time open charcoal stoves were placed, according to M. Soyer's directions, for the cooking of extra diets, heating coffee, &c.

In the Barrack hospital the kitchen was at one end of the yard, with no other arrangement in use than the above-mentioned boilers, which were, however, perfectly well set, and supplied with wood placed in fire-places, the opening of which was on the outside. The kitchen was twenty yards from one end of the barrack yard, and 200 yards from the other; the barrack yard being 220 by 194 yards; the corridor, therefore, which was in the internal side of the building, was about half a mile in extent. External to it were the wards, with their windows in the outside of the building. The mass of the patients was ranged on one flat and part of two other flats. In the winter of 1854–5, they varied from 1900 to 2500. The corridors had two rows of beds, placed transversely, leaving a vacant space along the centre of about three feet. The beds, measured across, with the intervals between them, extended over a space of from three to four miles, including, of course, both wards and corridors. As the eight coppers above mentioned, placed in one kitchen, had to supply three meals a day for this vast number, the issue was necessarily but slowly conducted—it occupied from an hour to an hour and a half. The meat was issued from the purveyors' store, at one office only, nominally between 9 o'clock and half-past 12, but the last of the meat was rarely brought into the kitchen before half-past 1, p.m. As soon as it was received by the orderlies, each carried the portion intended for his own mess or ward to the kitchen, and placed it in the copper, with some distinguishing mark (occasionally of very peculiar description), which was called a 'tally.' The nominal hour for dinner was one o'clock, when the portions were supposed to be cooked, and were re-delivered to the orderlies; and as no order of precedence was observed among them, it might thus easily happen that some of the portions remained three hours, and some not above half an hour in the coppers. The meat, after it had traversed the corridors, was divided into the required number of diets by the orderly, upon his own bed, and then distributed to the patients. The real hour of dinner was 2, 3, or even half-past 3, p.m.

No attempt at extra-diets had been made except that of cooking arrowroot in one copper and fowls and chicken broth in another, or in tin pots on Turkish braziers. The orderlies cooked some small articles for the patients in their own mess-tins.

Under these circumstances Miss Nightingale established an extra-diet kitchen, in November, 1854, which continued in action up to July, 1855. Four cooks were employed in it, besides which, arrowroot, puddings, lemonade, &c., were prepared during the whole time in the nurses' kitchen, for the patients.

The regulations for military hospitals contain five diets, being 'full,' 'half,' 'low,' 'milk,' and 'spoon.' When this last diet is ordered, consisting of 1 pint of tea and 4 oz. of bread for breakfast, and the same for tea, any articles may be ordered at the discretion of the medical officer, for the patient's other meals—these articles are then called 'extras.'

Under this regulation the medical officers drew for patients upon spoon diet (daily), 'extras' on diet-rolls for 700 to 800 men upon this kitchen.

The extra-diet rolls were necessarily carried away with the diets, to ensure a proper delivery by the orderly, and were returned to the ward for the medical officer to form his diet-roll next day.

In making the issues, therefore, the superintendent had only to comply with the diet-rolls presented, which did not return to her hands. The following extract from a parliamentary document will exhibit the ordinary demand.

Average Daily Issue of Extra-diets supplied from F. Nightingale's Kitchens to the Extra-diet Rolls of the Medical Officers, Barrack Hospital, Scutari, from the 13th January to 13th February, 1855. Vide p. 41 of 'Report upon the Scutari and Crimean Hospitals,' by the Cumming-Maxwell Commission.

No. supplied.	—	From Public Stores.	From Private Sources.
25 gallons . . .	Beef tea.	80 lbs. beef.	—
15 ,, . . .	Chicken Broth.	28 chickens.	12 chickens.
40 ,, . . .	Arrowroot.	—	Arrowroot.
15 ,, . . .	Sago.	—	Sago.
240 quarts . . .	Barley water.	Barley.	—
10 ,, . . .	Rice water.	Rice.	—
8 ,, . . .	Lemonade.	—	Lemons.
30 ,, . . .	Milk.	—	Milk.
275 portions . .	Rice puddings.	Rice.	—
15 bottles . . .	Port wine.	—	Port wine.
3 ,, . . .	Marsala.	—	Marsala.
3 ,, . . .	Brandy.	—	Brandy.
15 lbs. . . .	Jelly.	—	Isinglass.
4 dozen . . .	Eggs.	—	Eggs.
40	Chickens.	28 chickens.	12 chickens.

The average 'daily issue table,' though a fair average for the three months of December, 1854, January, 1855, and February, 1855, by no means reaches the amount of the issues of the end of December and beginning of January. For instance, in the return of issues for articles consumed in extra-diets, from December 19, 1854, to December 31, 1854, in one division (viz. D) alone of the Barrack hospital, containing 519 sick, appears an issue of 108 bottles of port wine mixed in the daily arrowroot from Miss Nightingale's kitchen.

A second extra-diet kitchen had been constructed by her in aid of the first about Christmas, 1854, and three supplementary boilers for supplying hot water for arrowroot, &c., were also set upon one of the staircases at her request.

A third of these kitchens was afterwards added by a first-class staff-surgeon, in charge of one of the four divisions of the hospital.

At the General hospital, in consequence of the difficulties said to exist, a modified system was adopted, until the extra-diets began to be cooked in the general kitchen arranged with M. Soyer in April, 1855.

M. Soyer was also permitted to arrange extra-diet kitchens at Haidar Pacha and at Koulali.

At the five general hospitals attended by female nurses in the Crimea, extra-diet kitchens, in which they were for the most part the cooks, were introduced under the female superintendent.

Butter being a new and exceptional article, was of course supplied from private sources when ordered by requisition of the medical officer. A quantity was despatched by her Majesty, and was placed by the purveyor in the charge of the superintendent, to be disposed of in conformity with the above regulation.

In consequence of the absence of any provision for sick and wounded officers in hospital, by the War Office regulations, the cooking for their service, in the Castle and General hospitals in the Crimea, was entirely done by the female staff; part also of the materials for their use being supplied from private sources. And, by the express desire of the medical officer in charge, the superintendent drew for the sick officers wine, bread, and meat, under the head of nurses' supplies, although not consumed by the nurses.

NOTE D.

Number of Shirts found in Patients' Knapsacks that are deposited in the Pack Store of the Barrack Hospital at Scutari.

Twenty-two.

SELKIRK STUART,
Purveyor to the Forces.

4th December, 1854.

Return of the Number of Shirts issued from the Purveyor's Stores in the General Hospital and Barrack Hospital at Scutari, from 1st October, 1854, to 16th February, 1855.

Issued.	To General Hospital.	To Barrack Hospital.	To Kulali.	To Haidar Pasha.	To Miss Nightingale.	Total.
From store at General Hospital . .	4,203	—	—	512	1,000	5,715
From store at Barrack Hospital . .	—	3,019	558	442	1,500	5,519
Totals issued .	4,203	3,019	558	954	2,500	11,234

Barrack Hospital, Scutari,
22nd February, 1855.

SELKIRK STUART,
Purveyor of the Forces.

Account of Clothing received into Quartermaster's Stores at Scutari.

Date.	Shirts.	Drawers.	Socks.	Mits.	Trowsers.	Boots.	—
1854. Dec. 5	589	1,173	4,628	—	—	—	Received from Constantinople, purchased by Captain Wetherall.
„ 16	3,588	1,817	4,597	—	—	—	
1855. Jan. 14	3,092	—	—	—	—	1,600	From Quartermaster-General.
„ 26	—	—	—	—	600	—	Purchased by order of Lord William Paulet.
Feb. 2	—	—	2,081	4,086	—	—	From Constantinople, purchased by Captain Wetherall.
„ 3	—	1,000	—	—	—	—	Purchased by order of Lord William Paulet.
„ 5	2,000	4,000	—	—	—	360	From Quartermaster-General.
Total	9,269	7,990	11,306	4,086	600	1,960	

Of the above clothing there has been served out to the men of the General Depôt, and to invalids proceeding to England, since the 5th December, as follows:—

Date.	Shirts.	Drawers.	Socks.	Mits.	Trowsers.	Boots.	—
	4,387	3,088	6,703	1,500	300	1,930	and 1,530 blankets.

JASPER HALL,
Captain 4th K. O. Regiment, Qr. Mr.

Scutari, 15th February, 1855.

The store was established when the first articles of clothing were received, viz., on the 5th December, 1854, as above.

JASPER HALL,
Captain 4th Regiment, Qr. Mr.

MAJOR SILLERY.

'I was commandant from the time the army left till within ten or twelve days. When convalescents or invalids leave the hospital, they come under my command. *Many of the sick and wounded men arrived with little or no clothing. From the want of any establishment for the purpose at this depôt, there is the greatest difficulty in supplying such men with necessaries.* There is a non-commissioned officer of each regiment here in charge of the men of his own regiment. It is the duty of that non-commissioned officer to meet the wants of the men if possible, getting the money for the purpose from the paymaster, who stops the amount from the soldiers' pay. The corporal must get the shirts when he can. In the case of boots, which are a heavy article, there is more difficulty. We cannot get the regimental boots here. For men going up to the Crimea, we look very closely as to boots. Till the last draft we sent up about a fortnight ago, we generally got boots from the commissariat for men going up; but I do not know if we got any for invalids. Every man is examined before he goes to the Crimea or home, but not when he comes out of hospital. We endeavour to complete the outfits as much as we can. This is done partly out of commissariat stores and dead men's effects. In the same way we give the red coatees of dead men.

'We want a quartermaster's establishment,—a large store with necessaries of all kinds. The complication of accounts with so many soldiers of different regiments requires a large staff. In a regiment, a soldier who wants anything is supplied by his captain, who inspects him and draws the articles wanting from the quartermaster's stores. Here we have no officer who discharges the duty of a captain.'

NOTE E.

Orderlies' Attendance.—With regard to the present 'regulation number of orderlies, viz., 1 to every 10 patients, it is to be observed,—

(1.) *Forty-bed Ward Minimum Size for Regulation Number of* 1 *Attendant to* 10 *Patients.*—A ward of 40 patients might be efficiently served (but it would be hard work) with

1 Head Nurse—Female.
3 Orderlies.

Provided always there were lifts and hot and cold water laid on.

With no number under 40 of patients to a ward, can the Regulation proportion of 1 attendant to 10 patients be adhered to.

(2.) *Twenty-bed Ward requires* 3½ *Attendants.*—A ward of 20 patients cannot be efficiently served (if the orderlies be men) with less than

½ Head Nurse—Female.
3 Orderlies.

And the other ward of this head-nurse ought to be on the same floor.

N.B.—The same number would quite as efficiently serve a ward of 30 patients, provided there be lifts and a supply of hot and cold water all over the building.

(3.) *Ten-bed Ward cannot be served by* 1 *Orderly.*—The Army system of 1 orderly to 10 patients, with a number not exceeding 10 patients to a ward, is upset as immediately by one bad case among the 10, as by 9 to the 10.

For, is the same orderly to be on duty for the 24 hours?

The difficulty is practically got over by the Army, with a permission that any 'bad case' may select any one he likes of his comrades (out of the depôt) to be 'told off,' to attend upon him.

This extraordinary regulation is equivalent to (and affords little other practical result, than) granting opportunity for any quantity of spirits, and illicit food, to be smuggled into hospital, and it is clear that it would be totally inadmissible in a general hospital, where the whole system of nursing would be under the most stringent discipline and supervision.

(4.) *Naval Hospitals Regulation number of Attendants* 1 *to* 7 *Patients.*—In all naval hospitals, the regulation number of attendants is 1 to every 7 patients, or 2 attendants for each ward containing more than 7 patients and up to 14.

In civil hospitals the proportion is as great, generally, of attendants to patients, and is mainly determined by the size of the ward:

E. g., in one hospital, where there are quadruple wards of 44 or 48 patients, 11 or 12 in each compartment, the number of attendants is 7.

In exceptional cases extra night-nurses, sometimes extra day-nurses serve particular patients. The labour, both of cleaning and of night-nursing, is much increased by the compartments being four, and separated by a large lobby.

In another of the large London hospitals,* where there are to each ward,

PATIENTS.		ATTENDANTS.
22 / 24	there are	1 Sister. 2 Nurses.
30	"	1 Sister. 2 Nurses. 1 Scrubber.
34	"	1 Sister. 3 Nurses.
40	"	1 Sister. 3 Nurses. 1 Helper.

In the Lariboisière Hospital at Paris, where the wards hold 32 beds, 1 sister, 1 nurse, and 2 orderlies on the men's side, 1 sister, 2 nurses, and 1 orderly on the female side, serve the ward efficiently. In this hospital there are no lifts.

(5.) *Same number of Men will not do same amount of Work as an equal number of Women would.*—One woman does the work of more than a man in a hospital, speaking of the duties discharged by under-nurses in civil hospitals; for men are not accustomed to these duties in England, as women are from their childhood.

From this it is by no means to be inferred that women of the class of under-nurses in civil hospitals should be employed in military hospitals, which unquestionably they should not. But it is to be inferred that the work will not be done efficiently, with a smaller number of men than would be employed of women.

(6.) Practically, it is impossible to serve 4 wards of 9 beds each, with

1 Head Nurse,
4 Orderlies.

For, as has been said, one bad case in each ward makes this economy as unmanageable as nine.

1 Female Head Nurse } to { 50 Patients, in (say)
6 Orderlies . . . . } to { 6 Wards,

would be wholly insufficient, though this attendance would be more than sufficient for 50 cases in one ward; but such a ward is considered

* It is singular how little, even in civil hospitals, attention has been directed to the comparative cost of nursing in larger and smaller wards. In two civil hospitals, the distribution of sick in which is nearly as in the two instances above, the annual cost of nursing each bed is about one-third *more* in the former than in the latter case. It is true that the average number of constantly occupied beds is about one-third less in the former than in the latter hospital. But the difference of cost seems mainly attributable to the difference of the number of beds in each ward. And the efficiency of the nursing is certainly not less in the latter than in the former hospital.

in a sanitary sense too large. Two wards of 30 beds each on the same floor would be efficiently served by such a staff, however; and there would be no sanitary objection.

(7.) *One Orderly should be the Frotteur.*—One orderly should be trained to be the *frotteur* to each ward. He should also be the porter to fetch and carry everything to and from the ward.

(8.) *Comparison of Cost of Nursing with larger and smaller Wards.*—The plan of Netley, with its wards for 9 sick, is by far the costliest for administration, as the following facts will prove:

I. It is proposed to provide the hospital with orderlies and nurses to conduct the nursing in wards of 9 sick, as mentioned.

II. On sanitary grounds wards may safely be large enough to accommodate 25 to 30 sick.

We may therefore choose the larger wards, being guided only by the cost of the nursing.

III. A ward of 9 sick would require 1 day and 1 night orderly, and a-third of a nurse (that is, a nurse could superintend three such wards).

A ward of 30 sick would require 2 day and 1 night orderlies and 1 nurse = 4 persons in all.

Or if two such wards were on one floor, 1 nurse could serve both.

IV. We cannot count the cost of orderlies and nurses, including lodging, rations, wages, at less than 50*l.* a year, which when capitalized at 3 per cent. (33 years' purchase), would amount to 1650*l.* for each.

V. A ward of 9 sick would cost in nursing 1650*l.* $\times 2\frac{1}{3} =$ 3850*l.*, or 427*l.* 15*s.* 6*d.* per bed.

VI. A ward of 30 sick would cost for nursing, in perpetuity, 1650*l.* $\times$ 4 = 6600*l.* = 220*l.* per bed.

[One nurse to each ward is here allowed.]

VII. The cost of the two plans relatively for a hospital of 1000 sick would stand thus:—

Wards with 9 beds =	£427,775
Wards with 30 beds =	220,000
Capitalized difference of cost in favour of large wards	£207,775

Suppose the sanitary requirement of 25 sick to a ward be combined with the greatest economy of administration, the cost would stand thus:—

For each ward of 25 sick, 3 orderlies, at 1,650*l.* =	£4,950
If two such are built in line close to each other, with the nurse's room between them, one nurse could superintend both wards, or half a nurse to a ward. The cost would be for the ward	825
	£5,775

Or cost for each bed $\frac{5775}{25} = £231$

The comparative cost of wards with 9 beds and 25 beds, would stand thus for 1000 sick:—

Wards with 9 beds	£427,775
Wards with 25 beds	231,000
Saving	£196,775

The cost of the administration per 1000 beds at Netley and at the proposed hospital at Aldershot would stand as follows:—

Netley	£427,775
Aldershot, pavilions, with 3 superimposed wards and 25 sick in each, would require 3 orderlies and 1 nurse* to each ward, and would cost 264*l.* per bed in perpetuity, or per 1000 sick	264,000
Difference of cost in favour of Aldershot . .	£163,775

Some abatement would have to be made, as regards the cost of Netley, as there are a few wards with 16 or 18 sick.

---

* One nurse might possibly be able to serve the whole pavilion. The highest estimate is here taken.

Note F.

The Director-General in 1858 states the Admissions and Deaths in the General Hospitals of the East, 1854 to 1856, thus:—

	SCUTARI.		KOULALI. Feb., 1855—June, 1855.	
	Admissions.	*Deaths.*	*Admissions.*	*Deaths.*
June, 1854 . . .	631	6		
July, ,, ...	267	13		
August, ,, ...	359	19		
September, ,, ...	3520	112		
October, ,, ...	1401	235		
November, ,, ...	3864	320		
December, ,, ...	3814	601		
January, 1855 . . .	4761	1393	Included under Scutari.	
February, ,, ...	1894	1084	794	302
March, ,, ...	2385	421	448	134
April, ,, ...	1629	149	138	52
May, ,, ...	1623	79	255	16
June, ,, ...	1519	41	328	5
July, ,, ...	2473	63	Not distinguished from Scutari.	
August, ,, ...	2981	58		
September, ,, ...	2195	46		
October, ,, ...	1187	44		
November, ,, ...	1124	163		
December, ,, ...	724	29		
January, 1856 . . .	448	20		
February, ,, ...	279	7		
March, ,, ...	595	8		
April, ,, ...	737	8		
May, ,, ...	586	1		
June, ,, ...	329	3		
	41,325	4923	1963	509
	Scutari and Koulali . . .		43,288[a]	5432[b]

[a] 4161 only from wounds.
[b] 395 only from wounds.

Note

The Director-General in 1858 states the Admissions and Dea

	VARNA. June, 1854—Jan., 1855.		GENERAL BALA Oct., 1854
	Admissions.	*Deaths.*	*Admissio*
June, 1854 . . .	201	4	
July, „	824	70	
August, „	445	99	
September, „	874	146	
October, „	268	23	512
November, „	150	26	514
December, „	57	1	598
January, 1855 . . .	27	5	752
February, „			421
March, „			295
April, „			164
May, „			324
June, „			230
July, „			192
August, „			194
September, „			218
October, „			191
November, „			152
December, „			102
January, 1856 . . .			82
February, „			60
March, „			105
April, „			54
May, „			186
June, „			341
	2846[c]	374[d]	5686[e]

[c] 197 from cholera.
[d] 148 from cholera.

[e] 333 fr
[f] 141 fr

he General Hospitals of the East, 1854 *to* 1856, *thus:—*

PITAL, VA.	CASTLE HOSPITAL, BALAKLAVA.		CAMP GENERAL HOSPITAL, CRIMEA.	
e, 1856.	March, 1855—June, 1856.		April, 1855—April, 1856.	
Deaths.	*Admissions.*	*Deaths.*	*Admissions.*	*Deaths.*
58				
60				
66				
50				
34				
37	218	3		
9	123	7	4	2
31	85	5	72	5
29	505	16	286	37
13	208	17	34	27
12	401	13	15	5
12	470	20	297	85
5	174	12	151	13
5	5	1	146	14
3	20	2	1	11
2	2	—	8	3
2	64	—	28	—
3	82	—	24	1
—	13	—	17	1
4	30	—		
3	154	—		
438[f]	2554[g]	96[h]	1083[i]	204[j]

…lera.
…lera.

[g] 1834 from wounds.
[h] 83 from wounds.

[i] 740 from wounds.
[j] 184 from wounds.

The Director-General in 1858 *states the Admissions and Dea*

	MONASTERY HOSPITAL, CRIMEA. July, 1855—June, 1856.		A Dec., 185
	Admissions.	*Deaths.*	*Admissio*
June, 1854 . . .			
July, " . . .			
August, " . . .			
September, " . . .			
October, " . . .			
November, " . . .			
December, " . . .			352
January, 1855 . . .			9
February, " . . .			117
March, " . . .			20
April, " . . .			226
May, " . . .			23
June, " . . .			29
July, " . . .	182	—	19
August, " . . .	104	10	18
September, " . . .	67	7	2
October, " . . .	115	6	
November, " . . .	40	2	
December, " . . .	48	1	
January, 1856 . . .	41	1	
February, " . . .	43	1	
March, " . . .	49	—	
April, " . . .	57	—	
May, " . . .	147	—	
June, " . . .	18	—	
	911	28	814

the General Hospitals of the East, 1854 to 1856, thus:—

OS.	SMYRNA.		RENKIOI.	
ept., 1855.	Feb., 1855—Nov., 1855.		Oct., 1855—June, 1856.	
Deaths.	*Admissions.*	*Deaths.*	*Admissions.*	*Deaths.*
3				
16				
14	737	30		
18	256	97		
13	94	9		
13	224	6		
4	38	5		
—	56	—		
—	26	1		
—	17	1		
	417	—	232	6
	22	5	234	1
			263	5
			345	12
			226	16
			11	9
			4	1
			8	—
			7	—
82	1887	154	1330	50

APPENDIX.

SITES AND CONSTRUCTION OF HOSPITALS.

(Three Articles reprinted from the 'Builder' of August 28th, and September 11th and 25th, 1858).

WE propose to devote two or three additional papers to a discussion of general principles on which hospital sites should be selected and hospitals constructed.

The sanitary history of hospitals may be summed up in very few words. There are hospitals on very bad sites: there are hospitals on comparatively good sites; but there is hardly an instance, in this country at least, of both hospital and site fully embodying those sanitary principles which are essentially necessary for a rapid recovery of the sick and maimed.

The one paramount sanitary condition which ought to be observed in all hospitals is that of having *pure and dry air* both within and without the walls of the buildings. The medical profession generally, and the public, have been so much in the habit of considering that medicine is some kind of occult force, by which disease is to be expelled from the human body, that it seems to have been thought sufficient for all purposes of curing and healing, that the sick man and the doctor should merely be brought together, in any locality, or under any conditions whatever. It used to be thought that human beings could live in any sort of place, and under any circumstances. And it is too much the fashion, even now, to consider that the sick ought to recover anywhere. Hospital trustees and committees conclude that they have done all that is requisite, when they have provided 'the very best professional advice and assistance that can be obtained,' and so the enormous mortality in hospitals has come to be considered as unavoidable, merely indicating the 'per-centage of mortality inevitably resulting from disease.'

It appears to us that sanitary reformers have too much overlooked the influence of hospital statistics upon the mortality of particular towns. It is true that the startling increase produced in the mortality of a registration sub-district by the accidental position of the union workhouse in it is well known. But it does not seem to us that the question has always been considered, whether much of the high mortality of particular towns is not due to the bad sanitary state of local charitable institutions. Towns, as everybody now knows, are more unhealthy than the country. But is it recognised that hospitals in towns follow the same law? That, other things being equal, a hospital in a town ought to yield a higher mortality, fewer permanent recoveries, a longer duration of sick cases, and therefore a greater current expense to the administration, than a hospital in the country would do for the same number of cases? Again, 'Hospitals,' says a great living statistical authority, 'are adapted to an intermediate state of civilization.' But the question of how much hospitals, in their present condition, increase the mortality among the poor of this 'intermediate state of civilization' has scarcely been touched. We can only state the question, and pass on. The importance of an inquiry as to what is the relative mortality in hospital practice and private practice among the poor (considering, too, that cases supposed incurable are never received into hospital) cannot well be overrated.

To return to our main question, we state broadly that 'town districts' should be avoided for hospital sites. Air of sufficient purity is not to be obtained in towns. Every existing town hospital ought therefore to be removed into the country, if it be possible to do so. At a moderate distance from towns land is much cheaper than in close-built places; and there are many large hospital establishments covering considerable areas of ground in crowded and valuable parts of towns and cities, which might be removed to the country, not only with incalculable advantage to the sick, but with great pecuniary gain to the hospital establishment. Even in so vast a place as the metropolis, a few casualty wards, where accidents might temporarily be seen, rooms for the examination and the reception of cases, and suitable vehicles for transferring them to the country, would be all that would be necessary to effect the reform.

The single consideration that the welfare and speedy recovery of the sick must be considered as the main object of all hospitals, ought to determine the removal and reconstruction of any unhealthy hospital, provided there be the pecuniary means of doing so; and if these are deficient, every exertion should be made to obtain them.

The foundations of every hospital should be laid in dry ground, which should be covered by concrete; or the building should have an arched basement, preventing the soil from having any communication with the floors above. Basements, enclosed and connected with the wards by staircases, are inadmissible, as affording ingress for damp air from the subsoil to the air breathed by the sick. A gravelly, or sandy subsoil, well drained, is the best. Clay subsoils are unsuitable in proportion to their power of retaining moisture. Made-ground, or ground covered with decaying organic matters, such as sweepings of streets, old burial-grounds, &c., should be sedulously avoided. One of the most ill-advised proposals for building a hospital which perhaps has ever been entertained is that of extending one of our great metropolitan hospitals over St. Clement Danes burial-ground, which has for years been a public nuisance to the metropolis.

The vicinity of marshes, wet ground, damp valleys, river estuaries, and mud-banks should be avoided. The choice of the position of the new Royal Victoria Hospital, at Netley, with its ten square miles of mud, is singularly unfortunate.

The natural configuration of the ground should admit of the buildings being protected from prevailing cold winds. The local climate should be dry. There should be a good supply of pure soft water.

Wherever 'Accident-wards' are required in towns, care ought to be taken to select a healthy district, not closely populated, with no nuisances nor other local causes of disease in the vicinity.

The great principle to be kept in view in the selection of sites and climates is that recovery from sickness in the vast majority of cases depends perhaps more upon pure air and pure water, with suitable diet, than upon any medical treatment, however skilful. The same remarks apply to surgical cases, in many of which there is a greater susceptibility to the influence of external causes than is to be found even in medical cases. The idea which ought to be uppermost in the minds of hospital committees, of architects, and of medical men, is that of *pure air*. Better leave the sick and maimed at home, unless this be realized. The aggregation of a number of helpless sick and maimed under one roof, in a given space, without attention to this requisite, may be at any time, as it has often been, and now is, nothing but manslaughter under the garb of benevolence.

The proof of this on a colossal scale, was the mortality at Scutari, which actually reached in the month of February, 1855, just before the Sanitary Commissioners commenced their labours, the incredible rate of 41.5 per cent. per annum. And this, when the character of the cases sent there from the Crimea had improved so much, that the general mortality of the army had diminished, notwithstanding up to this time the fearful and increasing death-rate of the Scutari hospitals.

In the Peninsula, as we learn from Sir J. McGrigor, the general hospitals had actually to be broken up, in consequence of the terrible rate of mortality within them.

Suppose a site already occupied by a hospital in which the condition above named, as first in importance, viz. that of *pure air* cannot be obtained. What is to be done? Ought we to go on receiving sick, in such a building, with the certainty that a number will be hurried prematurely to their graves year after year, who would have recovered, if the building had been in a more healthy place? Rather than admit so fearful an alternative, would it not be better to consider the question of removing the hospital altogether? Hospital sites in the midst of increasing populations, are far more valuable for almost any other purpose than for hospitals. Why not sell them, and remove the establishments into healthy districts, leaving only a few accident-wards, and offices for out-patients, and provide the means of transporting hospital patients to the new hospitals? Even medical schools ought to yield to true sanitary requirements. The sick ought not to be permitted to suffer, in order that any medical school, however good, may prosper. Medical schools would, however, prosper far more in the vicinity of good than of bad hospitals. The processes of nature by which the sick recover, would be seen in operation, instead of the processes of foul air by which the sick die, which latter, it is to be feared, forms a considerable proportion of the experience of some hospital students at the present time.

Having obtained a good site, how is that site to be used?—in other words, on what plan is the hospital to be built?

The fundamental idea of all hospital plans ought to be this: to have pure, fresh air in every part of the building. Fresh air is the *sine quâ non*. Unless a building can be so planned that the sick shall breathe air as fresh within its walls as they could do externally, they will suffer in a ratio corresponding to the degree of impurity.

All surgical operations, all medical treatment, and all nursing, are subsidiary to this great central point in hospital construction and management. We again assert that it should never be forgotten for a moment, by hospital committees, by architects, and by medical officers, that on the purity of the air of a ward depend, in a great measure, the recovery or death of the sick and maimed, the usefulness or injury arising from the hospital, the duration of cases; and, consequently, the hospital economy: whether, in short, a hospital planned, erected, and supported by 'voluntary contributions,' is to be a blessing or a curse to civilization. And when we consider that (in the words of the statistical authority already quoted), 'a man cannot forego his supply of air so many minutes as he can forego food for days,' is it a subject for wonder that pure air should be the main condition, as of health, so of recovery?

In the infancy of knowledge, when Christian benevolence provided for the sick and destitute, and where it also was the sole foundation of the work, hospitals were built in situations and on plans which were far from realizing the intentions of their founders. Mere shelter, food, and attendance were to be afforded to as large a number of sufferers as possible. In times of pestilence, the buildings would be crowded to excess, as we have seen in the case of the Irish workhouses during the famine of 1847. And who can tell how much of the dire loss of human life in the Middle Ages and during the great Irish calamity alluded to was due to benevolence misdirected.

One of the most striking illustrations of the results of absence of knowledge on these subjects is afforded by the experience of the Hôtel Dieu, at Paris. By the statutes of its foundation, 'all applicants' were to be 'admitted.' It had 1200 beds, and towards the end of the last century these beds used to receive at the same time from 2000 to 5000 sick. During epidemics as many as 7000 sick have been in the building at one time. From 20,000 to 30,000 sick passed through the hospital every year, about 25 per cent. of whom were carried to the cemeteries. In the other hospitals of Paris the mortality was about 12½ per cent. of the sick.

The excuse for the enormous mortality of the Hôtel Dieu was the same as that which is put forward by ill, and even by many well-informed persons at the present day for the high rate of mortality in civil hospitals, and, during war, in military hospitals, viz., that only the worst cases were sent there, and that they were sent there only to die.

The frightful over-crowding and bad ventilation, with the absence of every sanitary precaution, were, however, the real causes of the catastrophe in the Hôtel Dieu, just as the frightful over-crowding, the want of ventilation, defective draining, and want of cleanliness, were the real causes of the catastrophe at Scutari.

Both catastrophes led to great discussion and inquiry on the part of benevolent and intelligent men. In the Crimea the Sanitary Commission was the result; in the case of the Hôtel Dieu, it resulted eventually in the introduction into France of the greatest improvements in hospital construction and management which have taken place up to the present time. In these cases great good came out of great evil. From the Scutari case it is also hoped, that great permanent improvements will arise in military hospitals; and it will be well, if the experience of our own civil hospitals should lead to similar results.

Let any one conversant with the phenomena of disease go into a badly-constructed, and, consequently, ill-ventilated ward, and look at the sick. Unless his senses are dulled by perverse education, he will detect that peculiar musty smell which always indicates more danger to the sick, than there is safety contained in the long list of benevolent and eminent physicians and expert surgeons who attend in the wards.

Let him look at the languid powerless character of the patients, and ask himself the question, What would be the effect of fresh air upon these poor sufferers?

Let him ask if nurses, or doctors, or medical students are ever seized with fever; and he may elicit some facts which will astonish him.

Let him go into the surgical wards and ask whether wounds heal kindly, whether operations succeed, whether hospital gangrene ever appears, whether erysipelas is common, whether purulent ulcerations and discharges are apt to take place?

In the new surgical part of the Edinburgh Infirmary he will be answered that 'hospital gangrene is never out of the wards if full.' In the double wards of Guy's Hospital, in London, he will be told that they are only fit for medical cases. In the Scutari hospitals, he would have learned that out of 44 secondary amputations, 36, or upwards of 80 per cent. died; that in one month there have been recorded 80 cases of hospital gangrene!

To questions concerning the appearance of hospital fever among patients and attendants, the inquirer will also be answered too truly in the affirmative; and very likely he will be told at the same time, 'that it is nothing more than is to be expected, considering the kind of cases received into hospital.' This was the reply at Scutari, this is the general reply as to all casualties of the kind occurring in bad hospitals, civil and military. In fact, nature alone is to blame, according to these authorities, or else 'contagion' is the cause.

Nature is never to blame. If the cases be bad as possible, all the more necessity is there for care in placing them where they may have a moderate chance to recover.

To place patients in musty wards is simply to kill them, with the addition of torture.

The great army surgeon, Sir John Pringle, knew this quite well when he asserted that hospitals were amongst the chief causes of the mortality of armies. We may safely extend this remark, and say that badly constructed civil hospitals and other charitable institutions increase the mortality of districts.

We may take this for granted,—that no hospital ought to yield a mortality on its sick treated of seven to ten or eleven per cent., as is the case with our existing metropolitan hospitals. A certain per centage of deaths is inevitable, but not a per centage such as this.

Again, the whole doctrine of contagion, in the case of fevers, may be said to rest on no stronger foundation than the observation of facts in badly-constructed and ill-ventilated hospitals, where emanations from the sick play a corresponding part to the emanations from cesspools and other nuisances, in producing fevers out of hospitals. Both classes of emanations may become fatal or remain innocuous, but all depends on the extent to which they are diluted in pure atmospheric air. A few fever cases in a crowded, ill-ventilated ward may spread fever; but in a well-ventilated hospital, with plenty of cubic space, they certainly will not. The mortality in some of our best-constructed fever hospitals is enormous, and indicates of itself how much they stand in need of improvement.

When fever, erysipelas, or gangrene spreads in any hospital, such an event is no proof of 'contagion' or inevitable 'infection.' On the contrary, it is Nature's method of teaching men that her laws are being neglected. If a medical officer, or nurse, has fallen victim to the disease, we may be quite sure that the disease is not to blame, but those who failed to observe and to obey the laws by which the disease is infallibly prevented from extending itself.

Hospital sites and plans should be selected in strict conformity with Nature's laws; and not till this is done will the outcry against hospital contagions cease.

CONSTRUCTION OF HOSPITALS.—THE GROUND PLAN.

'HOSPITALS,' wrote an eminent French physician of the last century, 'are a curse to civilization.'

'Hospitals,' said Sir John Pringle, 'are among the chief causes of mortality in armies.'

It is not sickness, nor defective medical treatment, nor bungling surgical operations, which the great surgeon blames for the mortality. The medical treatment may be of the most perfect order, and yet be of little avail, if the buildings are bad or are overcrowded. Given a crowded, ill-ventilated hospital, and we are sure to have a high mortality; given the Hôtel Dieu, with its thousands of sick lying five and six in a bed, as was sometimes the case seventy or eighty years ago, and it is quite certain that a fourth part of all sick and wounded who enter such an hospital will be carried out of it to the cemeteries. Given an hospital for fever cases, in which there are four ranges of beds between the opposite windows, and no sufficient means to prevent the stagnation of air by changing it frequently, and we need not be surprised that, even in this temperate climate, we have a mortality of ten and eleven per cent. on the cases treated.*

Can we avoid by any structural arrangements such excess of mortality? This is the great question to be decided by hospital architects. Experience replies that there are many illustrations of masses of sick, who, having been treated in the open air, have escaped with few deaths; of large numbers treated in properly-constructed tents, and in small wooden huts, who have also recovered. Even the famine-stricken fever population of Ireland exhibited a marked contrast, in the small mortality of cases treated under hedges, in the open air, as compared with the same class of cases treated in workhouses and in hospitals. In like manner, in the large, but at first overcrowded, badly drained, and badly-ventilated hospitals at Scutari, the mortality rose as high as 42·7 per cent. on cases treated. In the wooden huts of the Castle Hospital above Balaklava, which only held from fifteen to thirty patients each, the mortality was under three per cent. of the cases treated. In the large overcrowded and badly-ventilated hospitals of the Peninsular army, the mortality was so great that they had to be closed, and the sick to be subdivided amongst a number of separate buildings. The old idea that fresh air was dangerous, and that warmth must be obtained and preserved even for small-pox and fever patients, has been done away with. A small-pox patient will have more chance of recovery if placed on clean straw under an open shed and covered with blankets, than in many private rooms, or in most hospital wards. Small-pox cases have been so treated, and without loss.

After the pestilential hospitals at Scutari had been cleansed, ventilated, and regulated according to the plans and under the directions of the sanitary commission, the deaths among the sick there did not much exceed the deaths among healthy guardsmen at home.

The lesson taught by all experience is, therefore, that large hospitals, as

---

* In our last section we gave the mortality of hospitals as 7 to 10 or 11 per cent. So stated, the mortality appears much less than it really is. The deaths on *cases treated*, may be fairly stated at 7, 10, and 11 per cent. When the element *per annum*, is introduced, it must refer of necessity to the number of beds constantly occupied by sick. And, if we apply this test to the hospitals whose deaths to cases treated we have given above, we find the *annual* mortality mount up to 85, 134, 110 per cent. This estimate, however, depends to a great extent on the rapidity with which the cases pass through hospital, as well as upon the mortality on the cases treated. The mortality on the cases treated in the metropolitan hospitals varies between four and a half per cent. and nearly sixteen per cent. Even this enormous mortality is a trifle when compared with the mortality in the metropolitan lunatic asylums. It varies from 9 to 10, 20, 25, up to nearly 42 per cent. of the cases which enter their destructive precincts. Surely there is 'something rotten in the state of Denmark.'

generally built and managed, are destructive of human life, unless extraordinary precautions be taken, and that it would be safer, as a general rule, in the absence of such precautions, to treat large numbers of sick in the open air. But we *must* have hospitals, and it is only by subdivision of the sick among a number of separate buildings, by relatively large cubic space, and by ample ventilation, that we can make hospitals furnish a minimum rate of mortality and a minimum duration of cases.

Large, rambling, low-roomed buildings, like old mansion-houses, are utterly unfit to receive sick. Their wards, or rather their badly-constructed, ill-arranged-rooms, are hardly adapted for the smallest families of healthy people to live in, much less for numbers of helpless sick people to recover in.

Given a building like Netley Hospital, with 500 sick confined to bed under one roof, in a series of cells communicating with each other by corridors and staircases, and we have not our own intelligence to thank if the sick are not carried off by hospital epidemics. There *is* a certain ratio between the number of sick placed in a building, and the amount of mortality. We know that, of the sick who were treated in the Crimea, almost exposed to the elements, during the frightful winter of 1854-55, not above one-half so many in proportion died, as perished in the great hospitals at Scutari during the same time. In the Crimea, the sick were aggregated, twenty or thirty together. In the hospitals at Scutari, there were, at one time, crowded under one roof, upwards of 4000 sick, wounded, and healthy men. The subdivision of the sick and wounded should therefore be made a primary object in all hospital construction.

To what extent should this subdivision be carried? We apprehend that the point must be determined by the twofold consideration of uniting the greatest advantages as to health, with the greatest facilities as to administration and economy.

And, first, we would lay down the principle that no hospital should be more than two flats in height. By such a construction, the sick are spread over a wider area, the walls are not so high as to interfere with sunlight and ventilation of neighbouring pavilions, the accumulation of hospital miasms in upper flats is avoided, access to the wards is easier for patients and attendants, and the whole administration is much facilitated. 'It will cost a great sum for land to build an hospital with only two flats,' will say the hospital economist. No doubt it will cost more money to accommodate a given number of sick in an hospital of two than in one of four flats. But the question has been discussed and decided notwithstanding, that the hospital of two flats is better than one of additional stories, and many hospitals of two flats have been built. Those of St. Jean and St. Pierre, at Brussels, have each two flats, and land is dear enough in Brussels. In this country also, land, in towns, has been used for building hospitals of two flats high. But even admitting the argument of the expensiveness of land as being entirely valid, the conclusion is certainly, not that hospitals three and four flats high should be built in towns; but that hospitals should be built in the country where land is less expensive. It is little else than a breach of trust to build great lofty architectural structures merely to flatter the bad taste of committees or governors; or to place the hospital in a close unhealthy neighbourhood, to suit the convenience of medical attendants, when the object of the whole ought to be, recovery of the sick.

The number of sick which may be safely placed under one roof will, to a certain extent, be determined by the local position of the hospital, by the amount of cubic space allotted to the patients, and by the state of ventilation. In the most recent hospitals the number varies considerably. Thus, in the

Royal Marine Hospital, which is at present being erected at Woolwich, 84 sick are to be accommodated under one roof. In the proposed hospital at Aldershott there will be arrangements for about 100. In each of the hospital blocks at Beaujon, in Paris, there are arrangements for 60 sick. In the magnificent Hospital Lariboisière, at Paris, the number is 102 per block. At the Hospital St. Jean, at Brussels, it is about 88.

Too great a subdivision of sick must necessarily incur an increase of cost in administration and nursing. On examining the experience of all these hospitals, we are of opinion that from 100 to 120 sick may be safely and economically treated under one roof, provided the ventilation and cubic space be sufficient, and the structure and communications of the building be so arranged as to facilitate the administration and nursing.

The next important question is, how to arrange buildings so as to have pure air outside and inside the structure.

The air must be 'moving air' in mass. The motion of the air in any room should never exceed a velocity of two feet and a half per minute, and should not be at any time much below this rate. There must be no 'stagnation.' It is a great error to imagine that because buildings are erected around a large court, therefore they are airy enough. A court, with high walls round it, does one thing with certainty,—it stagnates the air. All closed courts, narrow *culs-de-sac,* high adjacent walls, closed angles, overshadowing trees, and other obstructions to outer ventilation should be sedulously avoided, at whatever cost. Plan the building so that the sunlight can strike as large a surface of it as possible, and so that the air may move freely over the whole external surface.

Figures 1, 2, and 3 represent arrangements of buildings at present occupied for hospital purposes, which ought to be carefully avoided.

It may be considered certain that, wherever such arrangements exist, injury to the sick is so constant that, were it practicable, all the angles should be opened to admit of the circulation of air.

The simplest form of structure for ensuring light and ventilation is to build hospital wards in a straight line, Fig. 4, with windows on both sides, *i.e.*, back and front; the lengthway of the ward being the lengthway of the building, and the administration in the centre. By such an arrangement as this, however, no more than four wards could be obtained, if the building were two stories high. For small hospitals not exceeding 120 sick this plan would be economical and efficient. The direction of the axis of such a building should be from north to south, a little inclining to the east, which would ensure the sun shining on both sides every day of the year, and would also protect the wards from north-east winds.

One staircase would suffice for an hospital such as this. If it were carried from the bottom to the top of the building, and ventilated above the roof, it would cut off entirely one set of wards from the other, which is all that is necessary to prevent the possibility of any intermingling of foul air.

By adding projecting wings at the ends of such a line of building, as in Figs. 5 and 6, additional ward space might be obtained.

But additional staircases also there must then be at the ends, and such a building would have the disadvantage of a closed angle, although this would be of less consequence, if the wings were very short in proportion to the length of the front.

A much better arrangement is represented in Fig. 7, in which the wings are entirely detached from the centre, and connected with it only by an open

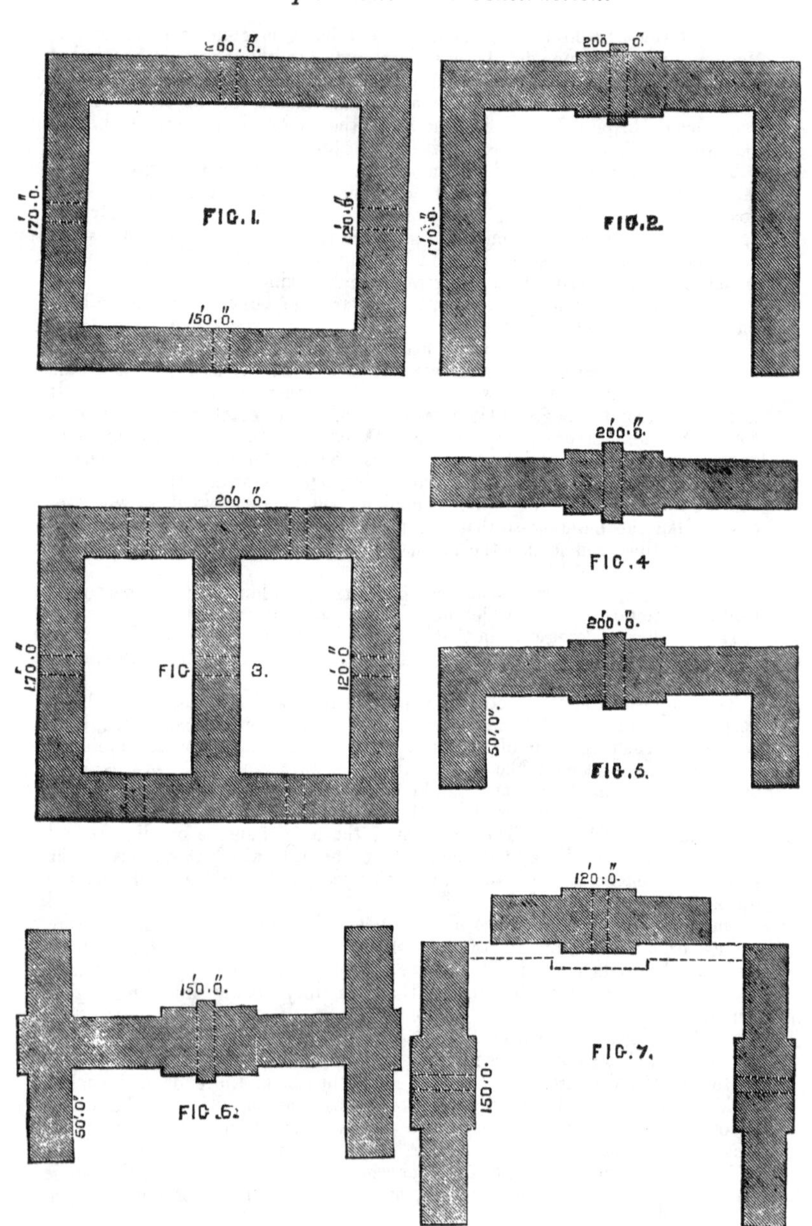

200. 0.
170.0.
FIG. 1.
120.0.
150. 0.
200 0.
170.0.
FIG. 2.
200. 0.
FIG. 4
200. 0.
170.0
FIG 3.
120.0
200. 0.
50.0.
FIG. 5.
120.0.
150.0.
50.0.
FIG. 6.
150.0.
FIG. 7.

corridor on the lower floor. This is the plan adopted in the great military hospital at Vincennes, and is a very good one for hospitals of a certain size, for the open angles permit air to circulate freely round the building. All of these plans, however, have the disadvantage of not admitting extension beyond a certain limit. The only plan which allows as much extension as can be necessary in any single hospital up to (say) 1000 sick (beyond which hospital management becomes very difficult), is the plan adopted in the hospital at Bordeaux; or still better, that of the Lariboisière at Paris. In that fine hospital, each block, containing 102 sick, constitutes a separate hospital. There are six of these blocks, which are arranged parallel to each other on two opposite sides of a square. And there are four blocks containing the administrative and other offices. The kind of arrangement is represented in Fig. 8.

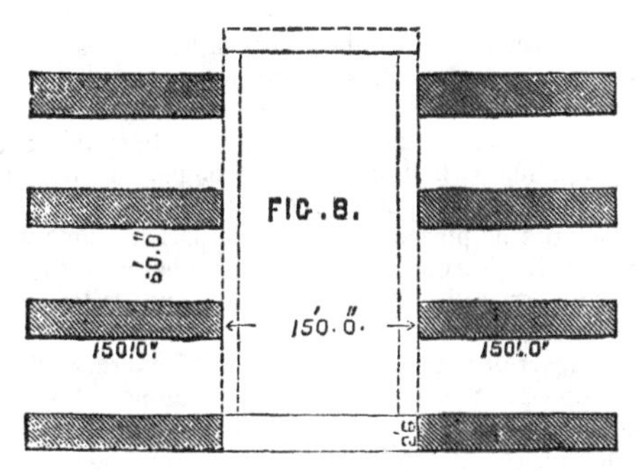

All the blocks are joined together by a glazed corridor along the lower flat, and by an open terrace above for convalescents taking exercise. In such a building, for the sake of sunlight, the axis of the wards should run nearly from north to south, and the distance of the blocks from each other should be about twice the height of the side walls.

We have given these illustrations of the arrangement of hospital buildings to show the more recent improvements in this important matter. Whether any of these or any better arrangement be adopted in building an hospital, it should be kept in mind that the great objects aimed at are, subdivision of the sick, free ventilation outside and inside the buildings, abundance of light, and windows on opposite sides of the wards, the heads of the beds being between the windows.

An hospital for 1000 sick ought to have none of its offices in duplicate. It should have one kitchen, so situated that the diets can be easily carried to any ward, and in the shortest space of time to all. The kitchen should be separately ventilated, and should not be under any part of the building used for sick. The laundry should be at a distance from the hospital, and in no way connected with it. The offices of the administration should be conveniently situated, not only for facility of access, but for efficiency of superintendence. This point was quite lost sight of in the Netley plans, in which, if the unfortunate governor happened to be wanted at the two extremities of the building successively, he would have to walk half a mile; while, with such a plan as that of Lariboisière, he could in the same time walk all round the buildings.

No sewer or drain should pass under any part of the hospital where there are sick. Sewers and drains should be outside, and free even of the external walls; there should be means of ventilation in abundance; there should be means of inspection and for flushing. The water-supply should be of the purest and softest description. It should be laid over the whole building hot and cold. This arrangement, together with that of lifts, saves the service of at least one attendant to every thirty sick. For water-supply, there should be a water tower and tank placed centrally, affording water under high pressure. The mains should be of cast iron, varnished; the service-pipes should be of wrought iron; there should not be any cisterns, but the supply should be constant. Fire-cocks, or 'hydrants,' should be placed at short intervals. Each supply-tap should have a waste-sink underneath it, and these should be placed so as to save labour. Water-closets must be external, but near the wards, and under such arrangement of plan as to prevent any possibility of atmospheric contamination. There should be not less than one closet to each ten patients. All staircases and passages should be wide, light, and airy. This is a most important consideration in hospitals. The steps should consist of stone, but they may be covered with wood, as in some of the French hospitals. Stone is essential to prevent accidents from fire. All hospitals should have convalescent wards, supplied with dining and day rooms. In the pavilion structure these can be placed under a separate roof. Casualty wards and small wards for special cases should be placed in a separate part of the hospital.

The square within the hospital, and the spaces between the pavilions, should be laid out as garden ground, with well-drained and rolled walks, and shaded seats for convalescents.

It is of great importance to provide places of exercise under shelter, for patients, to be appropriated to that purpose alone. Such recreation and winter-airing grounds may be comparatively large, and yet of cheap construction, if roofed on the Crystal Palace plan. The country can alone offer all the necessary facilities for the establishment of such valuable aids to hospitals.

HOSPITAL CONSTRUCTION.—WARDS.

In the two first sections on this subject, we have considered the question of Sites and general arrangement of hospital buildings. We now proceed to discuss the question of Ward Arrangements, with reference to light, ventilation, and administration.

Without entering into any scientific discussion as to the effects of light on organization, it may be taken as proved, that light exercises a peculiar influence on the two elements of form and colour. And, as both of these properties in living bodies are intimately connected with the *reparative* processes, we may safely consider it as certain that light is essential to the recovery of the maimed and sick. Experience in the more southern climates of Europe has determined the erection of houses for invalids in localities exposed to the full glare even of a southern sun, with which the rooms are flooded through a greater proportion of windows than would be considered safe by some of our medical men in colder climates. Very many of our civil hospitals would enable experiments to be instituted as to the injurious effects of absence of light on peculiar classes of disease. But the only experimental evidence we have hitherto obtained is that given by Sir James Wylie in regard to certain hospitals and barracks at St. Petersburg. In some of these hospitals there were rooms without direct light; and the sick and wounded treated in these dark chambers yielded only a fourth part of the recoveries when compared with patients treated in the light rooms. Nearly twice the number of invalids, it was also found, were received from the dark side of the barracks as from the light side. Dr. Edwards, who has paid particular attention to the effect of light upon health, has given equally strong testimony to its preventive and curative efficacy.

Altogether apart from scientific elements, common sense would lead us to anticipate such results.

Now let us see how light is treated by some popular physicians and ignorant nurses. In nine cases out of ten, a physician will draw down the window-blinds, and half shut the shutters, while an ignorant nurse will probably shut the remainder of the shutters—especially if it be a bright day—and draw the bed-curtains. We have the positive testimony of a well-known London physician, given in his report to the Netley committee, that whenever he enters a sick-room, he takes care that the bed shall be so placed that the patient shall be turned away from the light. (See note, p. 108). After this, we cannot blame army medical officers for not knowing much about the matter. An acquaintance of ours one day passing a barrack, saw the windows on the sunny side boarded up in a fashion peculiar to prisons and penitentiaries. He said to a friend who accompanied him, 'I was not aware that you had a penitentiary in this neighbourhood.' 'Oh!' said he, 'it is not a penitentiary—it is a military hospital. There is a great horror of light on the part of certain army medical men. I suppose,' he added, 'the medical officers are afraid the light will alter the shape of the men.' Not a few civil surgeons, also, treat light as if it were an enemy.

In the teeth of all these popular fallacies, we assert that every sick ward should be capable of being flooded by sunlight; and, consequently, that the windows should bear a large proportion to the wall-space in all hospitals. Experience appears to prove that window-space should not be in a much less proportion to wall-space of an hospital than one to two.

The next important question is, the quantity of air required in an hospital, and how best to renew it.

The great and constant movements going on in the atmosphere prove that the amount of change which nature has provided for healthy existence is unlimited. The test of the ventilation of a sick ward is the comparative freshness or impurity of the air. The interesting experience of Lariboisière appears to prove that about 4000 cubic feet per bed per hour are required to ensure this.

There are two ways of maintaining the freshness of the air of a ward:—

First, by so constructing a building that nature will renew the air, if left to herself, which is by far the best plan.

Secondly, by artificial ventilation—never to be used except as a *pis aller*.

If the hospital is badly planned, or if fuel be dear, artificial ventilation comes into beneficial operation; for it admits of economical warming, but it never *freshens* a ward like pure natural air from without. Every intelligent physician and nurse knows the value of freshness in the air of a ward. It is better oxygenated, and, perhaps, contains more ozone. But whether or not this be so, it is quite certain that a condition of ward-air is secured by open windows, and by open fire-places for warming, which is never obtained by the best ventilating machinery, especially if warm air be thrown in by it.

Architects must therefore well consider how in every corner of wards, passages, and staircases, air is to be kept constantly *flowing*,—not air passing in a stray or in a strong current, but air gently moving.

This can only be done by having wards of a certain height and breadth, and by having a window for every two beds, the windows opposite.

We shall afterwards discuss the questions as to the size of wards, and the proportion of window-space; but, in the meantime, we lay down a principle of construction applicable to all wards, of whatever size.

Few greater mistakes can be committed in hospital construction, as far as light and ventilation are concerned, than placing the windows at *one end* of a ward, or even at both ends, with beds ranged down the sides, their heads to the dead wall. No greater mistake, we had nearly said, could be committed. But there has been a greater in the case of Netley Hospital, where, not only has this most objectionable principle of construction been adopted, but also that of covering the windows in on one side by a glazed corridor.

In an hospital similarly built, we asked the medical officer how his sick, who were men in the prime of life, recovered? His reply was, 'I do not know how it is: but all my cases linger.' The reason why the cases lingered was obvious enough. For, in the first place, the air of all the wards communicated by the corridor; and, in the second place, wards and corridor were so arranged that thorough ventilation was impossible.

In another such hospital we expressed to the medical attendant our surprise that every ward was so uniformly full of sick. He replied that it did not arise so much from fresh admissions, as from the fact that when he got sick into the hospital, he could hardly get them well again.

Our military buildings have been and are most unfortunate in their arrangements. The unhappy Netley Hospital has been copied in its wards and corridor arrangements from such places as Chatham Garrison Hospital, Woolwich Hospital, &c. In fact, there are few military hospitals in which the error does not exist in one form or another. It was hardly to have been expected that, given the same directors, Netley should have escaped the same fate. A great mistake and a lamentable misfortune is Netley Hospital. But let us hope we have seen the last of such fatal blunders in hospital building.

Having alluded to military hospitals, we may say a word about regimental hospitals. We have visited several of these establishments, and certainly we can but consider their style of construction as extraordinary. They generally resemble a small ill-planned village residence, belonging, perhaps, to the attorney or the doctor. There are usually a narrow passage, a narrow staircase, and small rooms, in which the sick are stewed up: these rooms have a window or two here and there, as if each building had been an especial victim of the window-tax. Buildings of this class have not such a thing as a proper ward. They have merely little bedrooms, and everything else upon a little scale. As for light and ventilation arrangements, these have not attracted much, if any, attention on the part of the adapter. We hope that there may be some regimental hospitals on a better plan. We speak from a knowledge of such as we have seen; and, judging from these, we should say that a minimum of accommodation has been provided at a maximum of expense.

Besides arranging for light and ventilation, all hospital-plans should provide for the greatest economy in administration, consistent with healing of wounds and recovery of the sick. There should be facility of superintendence and nursing. The wards should be of such a size, and, if possible, so arranged, that the head nurse may have all her sick under her eye at once. This is especially necessary for night-watching. A single attendant can easily perform the duty of night-watching for many bad cases, if the beds are so placed as to be seen from a single point. The head nurse's room should have a window looking into her ward. It is highly important that patients, who must necessarily be in various stages of sickness or convalescence, should feel that they are continually under the eye of the head nurse. It is, of course, most economical to have one ward to each head nurse. The ward, therefore, should be large enough to occupy her whole attention; but not too large to render its ventilation difficult. Small wards cannot be overlooked with sufficient strictness, unless, indeed, a head nurse is to be set apart for each—an arrangement which would enormously increase the expenses of nursing, without benefiting the patients.

A head nurse may adequately superintend a ward of from thirty to forty patients. But, if we are to be guided by the results of recent experience in hospital building, we should say that a ward with thirty sick, or thereabouts, is, upon the whole, the best for sanitary reasons, and this number of thirty we propose to fix as the ward unit for an hospital.

Much has been said about the benefits of small wards for from six to ten sick, about the greater comfort and privacy of such wards, and the greater facility for ventilation which they afford. It is simply an error to assume that small wards afford any such advantages. Privacy, in an hospital, does not extend beyond any two adjacent beds. As regards complete ventilation, the effects of angles in retarding an even flow of air has not been sufficiently considered. Direct experiment, made in the wards of the Lariboisière Hospital, has proved that the amount of air circulating along the centre of a ward is two or three times as great as it is near angles. Such a result might have been inferred. But the important practical point never seems to have been comprehended, that the difficulty of ventilating a given cubic space occupied by sick bears a direct ratio to the length of the corridors and to the number of cells or 'wards' into which that space is divided. Unnecessary rooms, angles, or cupboards should be omitted in any structure. There should be no dark corners in any part of an hospital ward. Every recess or angle not easily overlooked is as injurious to hospital discipline as it is to hospital ventilation. Each pavilion should have a staircase, wide, roomy, well lighted, and ventilated from above. The gradients of the steps should not exceed a rise of 5 inches to 12 inches tread, to enable patients to ascend and descend with facility. There should be as little passage space as possible, and none of it should be dark. The head nurse's room should be situated, as already described, with one window opening down the ward, and a window looking into the open space. There must be a separate scullery for every ward, adjoining the nurse's room, but having a separate entrance and a separate window to the open air. A lift will be found an essential means of convenience as well as of economy. Each ward should have bath-room, lavatory, and closets. On the male side there may be other necessary conveniences attached. The best situations for these will be at the ends of the wards, opposite the entrance. They should be separated from the wards by a partition wall with a half glass door. There should be a small lobby, lighted by a large window at the end, and from this lobby there should be doors opening right and left—one to the bath-room and lavatory, the other to the closets, &c. The ventilation and lighting of these places should be ample and independent of that of the wards. No ward requires more offices than are here enumerated. (See note, p. 108.) There should neither be small wards nor dining-rooms attached to the large wards.

A ward for sick should not, as a rule, contain convalescents. This class of patients, under a proper system of hospital construction, should always have separate accommodation. Small wards for casualty cases, as already stated, should be built separately and be separately administered.

In applying these principles it is necessary, first, to ascertain what ought *not* to be done.

Figures 1, 2, 3, and 4 represent common errors in hospital ward construction, which ought to be carefully avoided. Figure 1 represents the arrangement in King's College Hospital and in the new wards at Guy's, London. Figure 2 is the ward arrangement usually carried out in buildings which have been adapted for civil hospitals. It was designedly adopted in the military hospital, Portsmouth, as also in many other military hospitals. Figure 3 is the arrangement in a few ill-constructed foreign hospitals, as *e.g.* in one at Rotterdam. It exists at Chatham Garrison Hospital, and is to be perpetuated at the new Victoria Hospital, at Netley. Figure 4 shows the kind of arrangement in the

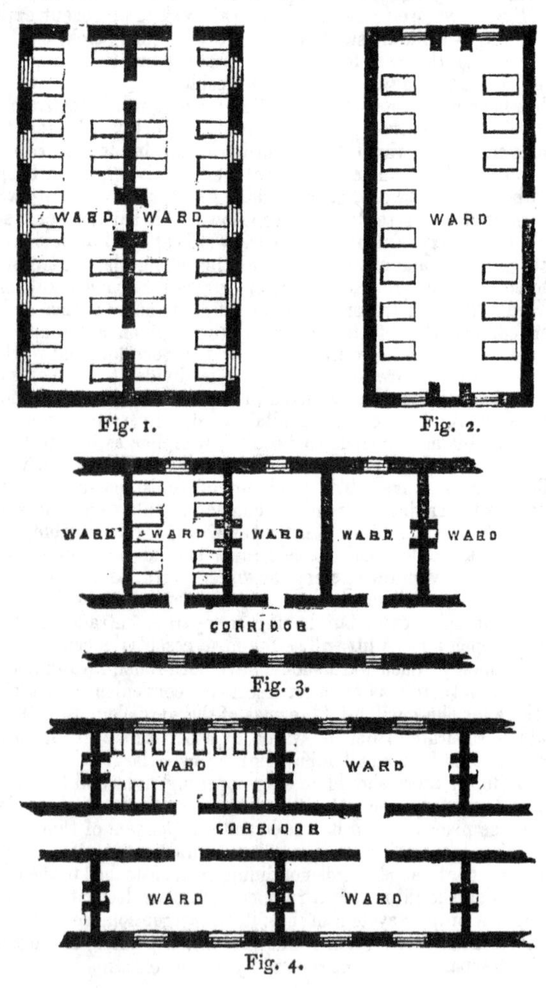

Fig. 1.

Fig. 2.

Fig. 3.

Fig. 4.

wards and corridors in the old Royal Marine Hospital at Woolwich. These diagrams do not pretend to extreme accuracy; they simply indicate the arrangement.

These plans should be avoided for the future in hospital construction. They are contrary to every sound principle of sanitary construction, and can only be kept free from hospital epidemics by the greatest possible care,—hardly even with that.

We shall next state what ought to be done in applying the principles of hospital construction we have laid down.

The best principle of hospital construction is, as has long been urged, that of separate pavilions placed side by side, or in line. The former is preferable for large hospitals, for the reasons stated in our last paper. The distance between the blocks should be double their height. There should be two flats in a pavilion, and one ward to a flat.

The hospital should be erected to form a square; the basement story of the pavilions being connected by a corridor, and the whole of the basements erected on arches.

The wards should hold from twenty to thirty sick: each bed should have from 1500 to 2000 cubic feet of air space allotted to it.

The following table exhibits the proportions of a ward for thirty-two patients. The first column gives the proportions of such a ward in the Lariboisière Hospital; the second, the proportions adapted to a larger cubical space, such as is given in our best hospitals in this country:—

	1.		2.	
	Feet.	In.	Feet.	In.
Length of ward	111	6	128	0
Breadth	30	0	30	0
Height	17	0	17	6
Wall spaces between end walls and windows	5	0	6	4
Breadth of windows	4	8	4	8
Breath of wall space between windows	9	2	11	4
Height of windows	13	0	13	6
Cubic space per bed	1760	0	2100	0

It is undesirable to increase the width of any ward beyond 30 feet, because the distance between the opposite windows becomes then too great for efficient ventilation.

The proportions of a ward for twenty patients might be 80 feet long, by 25 feet wide, and 16 feet high. This would give about 1600 cubic feet to each bed.

One window at least should be allotted to every two beds. There are hospitals with a window to each bed. The windows should be double, or be glazed with plate-glass, to prevent loss of heat. Tripartite windows, like those of Middlesex Hospital, are useful for ventilation. The ward walls should consist of pure white polished Parian cement, or some other equally white non-absorbent substance. Grey-coloured cements should be avoided: they never look clean; they give the ward a sombre appearance, and they hide dirt. The best ward flooring is oak. No sawdust nor other organic matter, capable of

DESIGN FOR A PAVILION HOSPITAL.

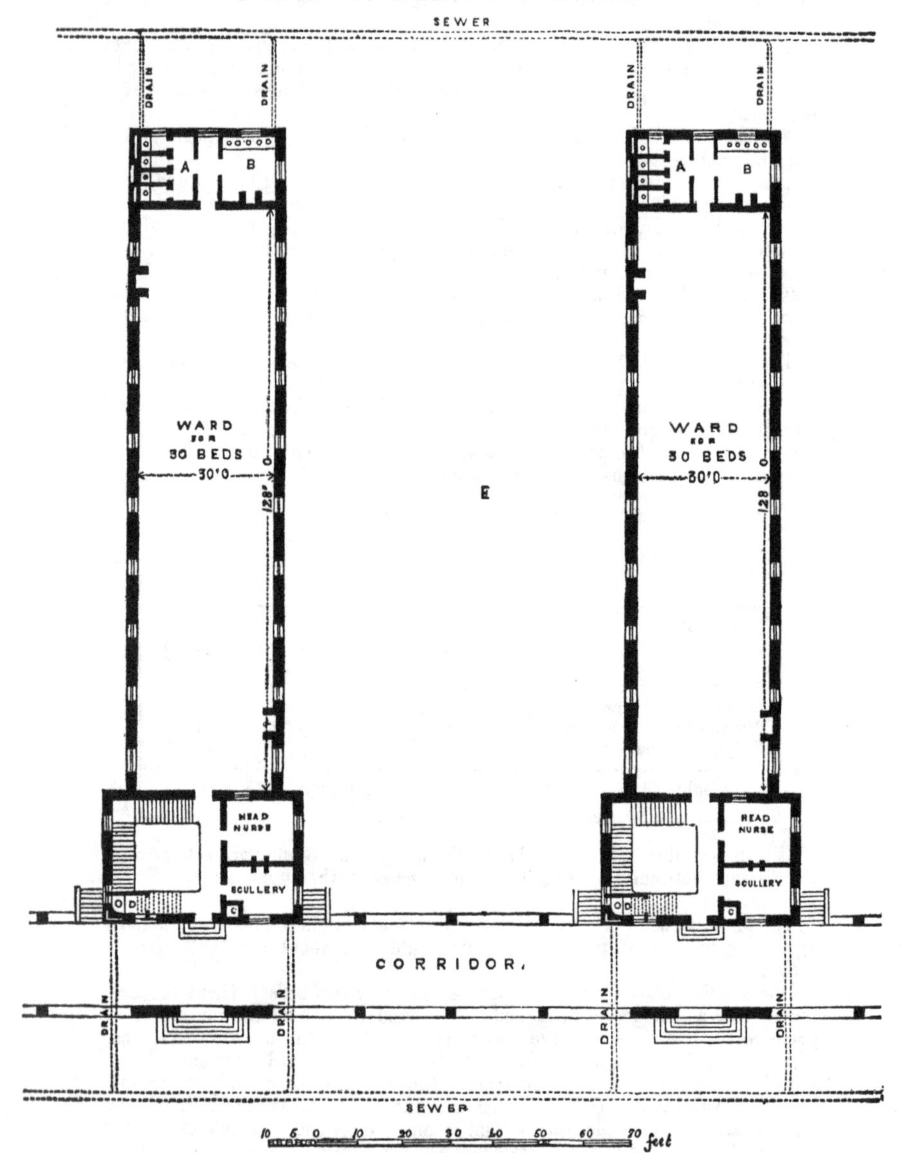

Fig. 5.

A. Ward Closets.
B. Bath and Lavatory.
C. Lift in Scullery.

D. Private Closet.
E. Ornamental Ground.
Ward Windows to be 4 ft. 8 in. in the clear.

rotting, should be placed underneath hospital floors. The joints of the flooring should be well fitted together, so as to be impervious. Floors should be bees-waxed, or oiled and polished. The general baths of the hospital should be separated from the pavilions, but connected with the corridor. They should contain hot and cold water, sulphureous, medicated, hot air, vapour, shower, and *douche* baths. The kitchen should have walls and ceiling of Parian cement.

There should, as stated, be a head-nurse's room and scullery attached to each ward, and store presses in each ward.

The bedsteads should be of iron. The ward furniture should be of oak. The arrangements for water supply and drainage were discussed in a previous article. We have also pointed out already the best position for closets, lavatories, ward-baths, &c.

The ward construction, now described, is that which, up to the present time, experience has shown to be best suited for fulfilling all the requirements of ventilation, light, cheerfulness, recovery of health, and economy, in this country. It is embodied in the plan, Fig. 5.

One great advantage of the proposed system of construction is, that it admits of any arrangement of the pavilions on plan, which is consistent with light and ventilation. Hospital establishments, so constructed, may be added to without difficulty and without altering, or indeed without interfering with any of the existing pavilion buildings.

The principles of hospital construction which we have now laid before our readers are a tacit condemnation of the majority of hospitals in the three kingdoms. It may be thought by some of our readers that we have expressed our opinions somewhat strongly. We have considered it necessary to do so, because, as we have shown, large hospitals are even now being built in defiance of the most ordinary sanitary principles, as to site and plan; and men have been found to defend these errors. Another opportunity is now afforded of retracing our steps. The noble bequest of Mr. Morley for the foundation of a convalescent establishment in the country, near London, may be a great practical advance in the right direction; and it behoves the governors of St. George's Hospital to take care that an institution of such importance is constructed so as to realize the benevolent intention of its founder.

Most of the great hospital and charitable institutions in this country were suburban at their foundation; the rapid growth of our towns in modern times has encroached so much upon space once country, that gardens and fields have been absorbed, and are now covered by bricks and mortar; and hospitals are surrounded by the screaming and roaring traffic of railways, street cabs, omnibuses, and wagons. It is due to the benevolent founders of our great charitable institutions, that their humane intentions should continue to be realized by removing the sick and maimed to pure air and quiet. It is due to poor suffering humanity that any plans adopted should be the most perfect modern intellect can devise; and it is no less due to the cause of charity that there should be the best and most economical form of hospital nursing and management.

The following is a summary of the principles of hospital construction contained in this and preceding papers:—

1. Never erect a general hospital within the precincts of a town, or in suburbs likely to be built upon.

2. Remove all large hospital establishments out of town, or from populous suburbs, as soon as possible.

3. Build all hospitals in the country, on areas of ground sufficient to admit

of extensions of the buildings, and to prevent other buildings from being erected within such a distance as shall interfere with a free circulation of air and with quiet.

4. Select a dry and mild climate.

5. Give the preference to a porous self-draining sub-soil, and avoid ground saturated with organic matter, especially old graveyards.

6. Build all hospitals on arches, to admit of a free circulation of air underneath the ward floors.

7. Let the plan be simple, and have as few closed angles and corners as possible.

8. Do not provide for more than 120 patients or beds, under one double pavilion roof. If the hospital must contain a larger number of beds, increase the number of pavilions, but on no account increase their size. Connect the pavilions by corridors running between the ends, and no higher than the ground-floor.

9. Plan any hospital with no more than two flats, containing, in a single pavilion, two superimposed wards. Provide for convalescents and 'casualty' cases in pavilions separate from the ordinary sick.

10. Provide for no more than 32 beds in a ward—16 on each side, with a window to every two beds.

11. Let the window spaces be, as near as may be, in the proportion of one to two of the wall-space. The windows should reach from within 3 feet of the floor to within 1 foot of the ceiling.

12. Wards should not exceed 30 feet in width.

13. Each bed should have from 1500 to 2000 cubic feet of air space.

In very exposed and airy situations the smaller cubic space will suffice; but where, from the nature of the ground, there is not likely to be much horizontal external movement of the atmosphere, the larger the internal cubic space is the better.

14. Trust for pavilion ventilation to open windows and fireplaces. Artificial methods are, in this climate, unnecessary, with proper construction.

15. Place water-closets, ward-baths, and lavatories, at the far end of a ward, opposite the entrance, and, in addition to ventilating them, cut them entirely off from the ward by a separately-ventilated and lighted lobby.

16. Restrict the ward offices to a nurse's room and scullery, with a lift.

17. Let staircases be wide, roomy, and thoroughly ventilated up to the roof. Construct the stairs and entrance-lobbies of stone, and cover them with wood.

18. Make hospital floors of oak, and walls and ceilings of pure white polished cement.

19. Supply hospitals with water at high pressure, and lay it on, hot and cold, over the whole of the buildings.

20. All sewers and drains must be outside, and detached from the walls of the buildings. Provide for their ventilation at a distance from the wards, and also for their inspection and flushing.

21. Provide garden ground for exercise with properly drained and gravelled walks, sheltered seats for convalescents, and, where practicable, a promenade covered with glass for use in bad weather.

Other matters of detail, regarding kitchens, washhouses, rooms for hospital administration, &c., can be best decided on according to local circumstances; but none of these should ever be under the same roof with the sick.

The public hospitals of any country may fairly be taken as a standard of the knowledge and care of the governing body, or of civilization amongst a people. Commercial wealth, political freedom, and education ought to enable the inhabitants of Great Britain to set an example in this respect. Life is of more value in the British empire than in any other part of the globe, and the means to secure it to the latest periods ought to be sought out and put in practice. Let us reform our hospitals now; and, as we improve our cities and towns, they will be less required.

Note to Page 100.

To show the want of acquaintance with the influence of light on recovery evinced by otherwise intelligent members of the medical profession, it is only necessary to quote the following passage from page 125 of the report on Netley Hospital, 1858: 'In the side chambers of private patients medical men generally enjoin that the bed should be placed in any other position rather than opposite a window.' We are glad to know that there are medical men who entertain far more just and enlightened views as to the influence of light on disease.

Note to Page 102.

It is important, for the purposes of discipline, to isolate as much as possible the nursing staff of each ward. The head nurse should be within reach and view of her ward both by day and night. Associating the assistant nurses in large dormitories tends to corrupt the good and make the bad worse. Accommodation for the assistant nurses of each ward separate should therefore be found in each pavilion. This may easily be done in the roof over the front part, and the great staircase of each pavilion. In doing this, care must be taken that the night nurse shall have quiet to sleep by day. She should have a room to herself.

THE END.

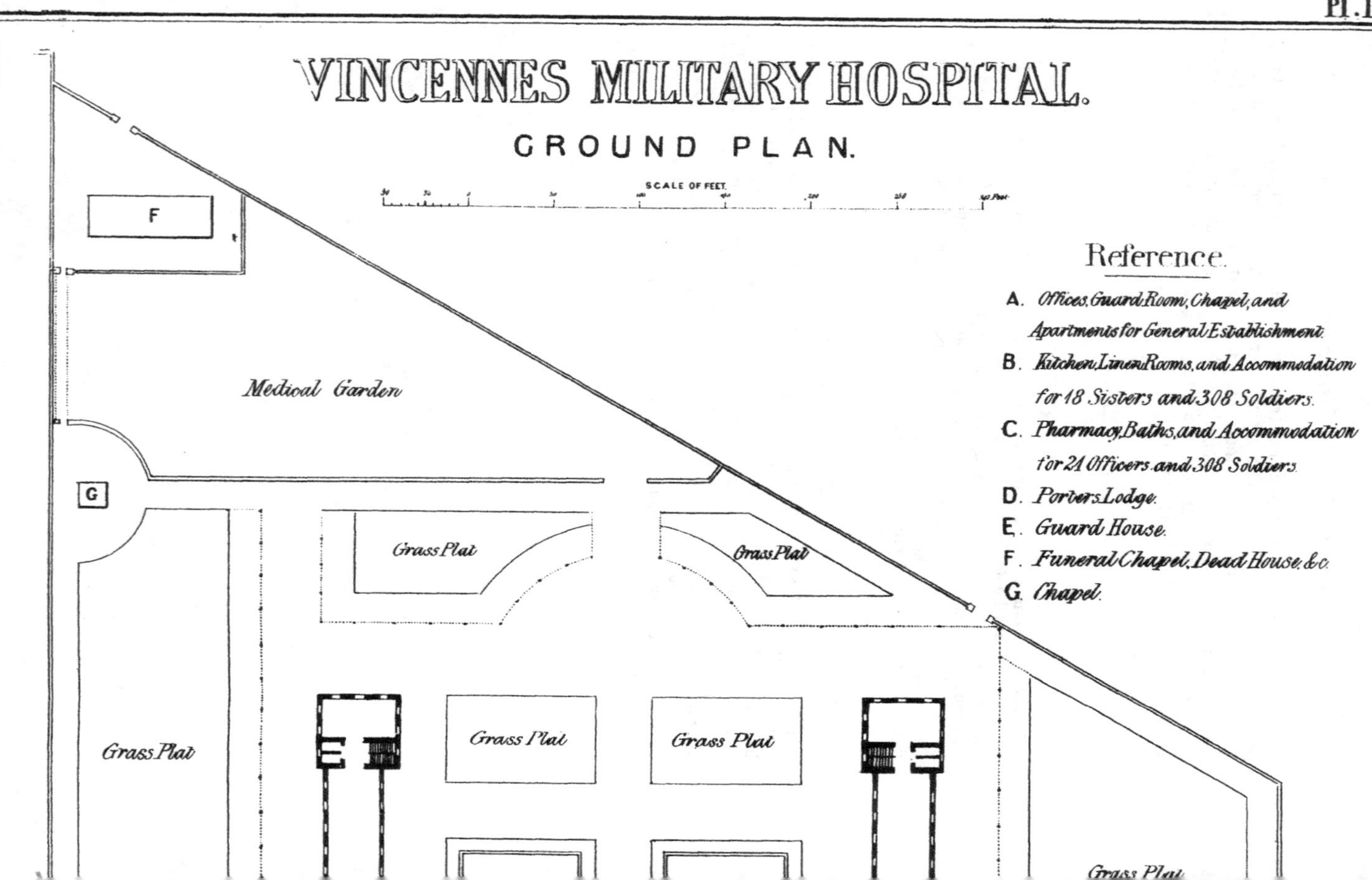
Pl. 1.
VINCENNES MILITARY HOSPITAL.
GROUND PLAN.
SCALE OF FEET.
Reference.
A. Offices, Guard Room, Chapel, and Apartments for General Establishment.
B. Kitchen, Linen Rooms, and Accommodation for 18 Sisters and 308 Soldiers.
C. Pharmacy, Baths, and Accommodation for 21 Officers and 308 Soldiers.
D. Porters Lodge.
E. Guard House.
F. Funeral Chapel, Dead House, &c.
G. Chapel.
F
G
Medical Garden
Grass Plat
Grass Plat
Grass Plat
Grass Plat
Grass Plat
Grass Plat

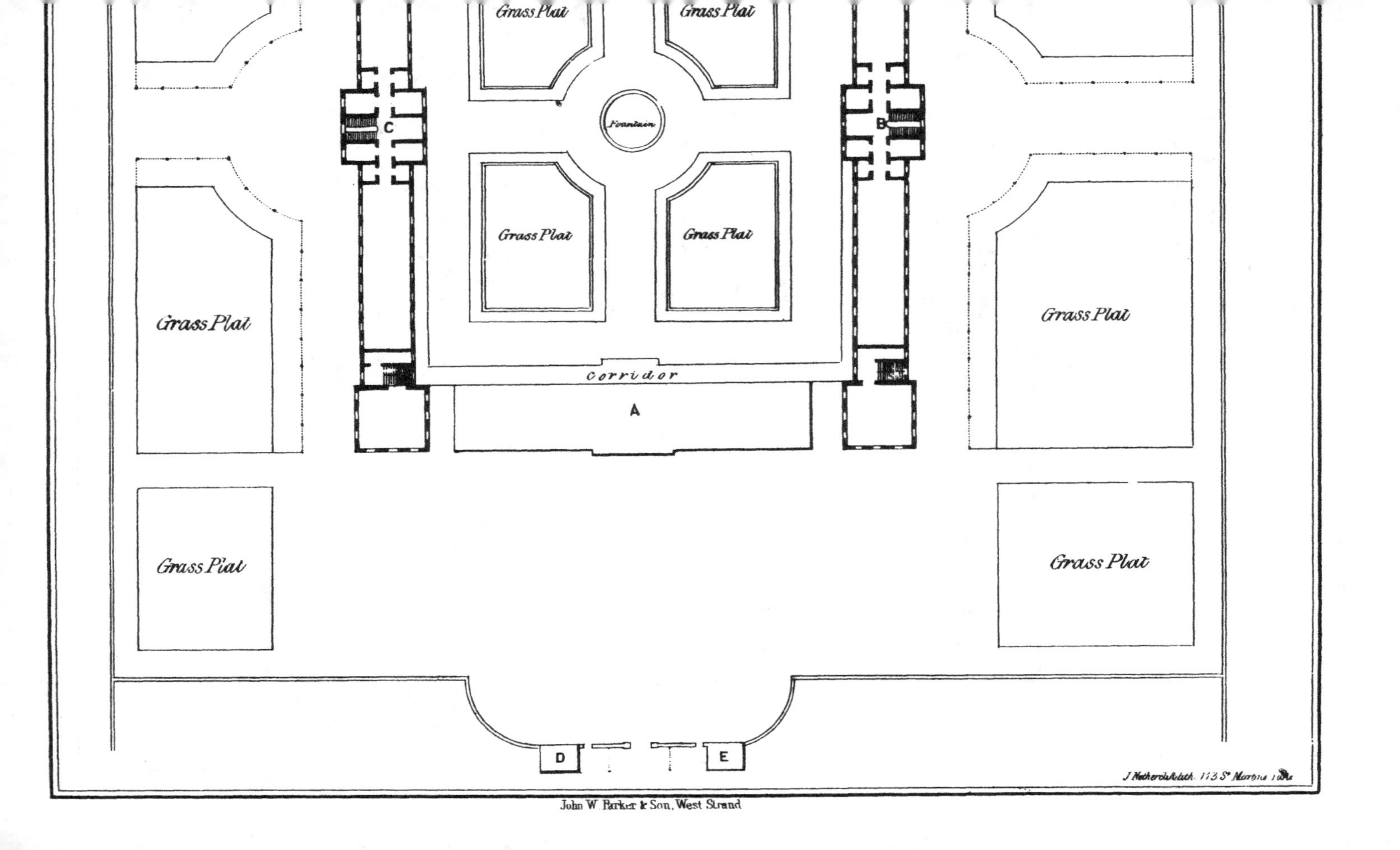

John W. Parker & Son, West Strand

GROUND PLAN OF THE

O

ROYAL VICTORIA

SHOWING ONE SIDE OF EACH W

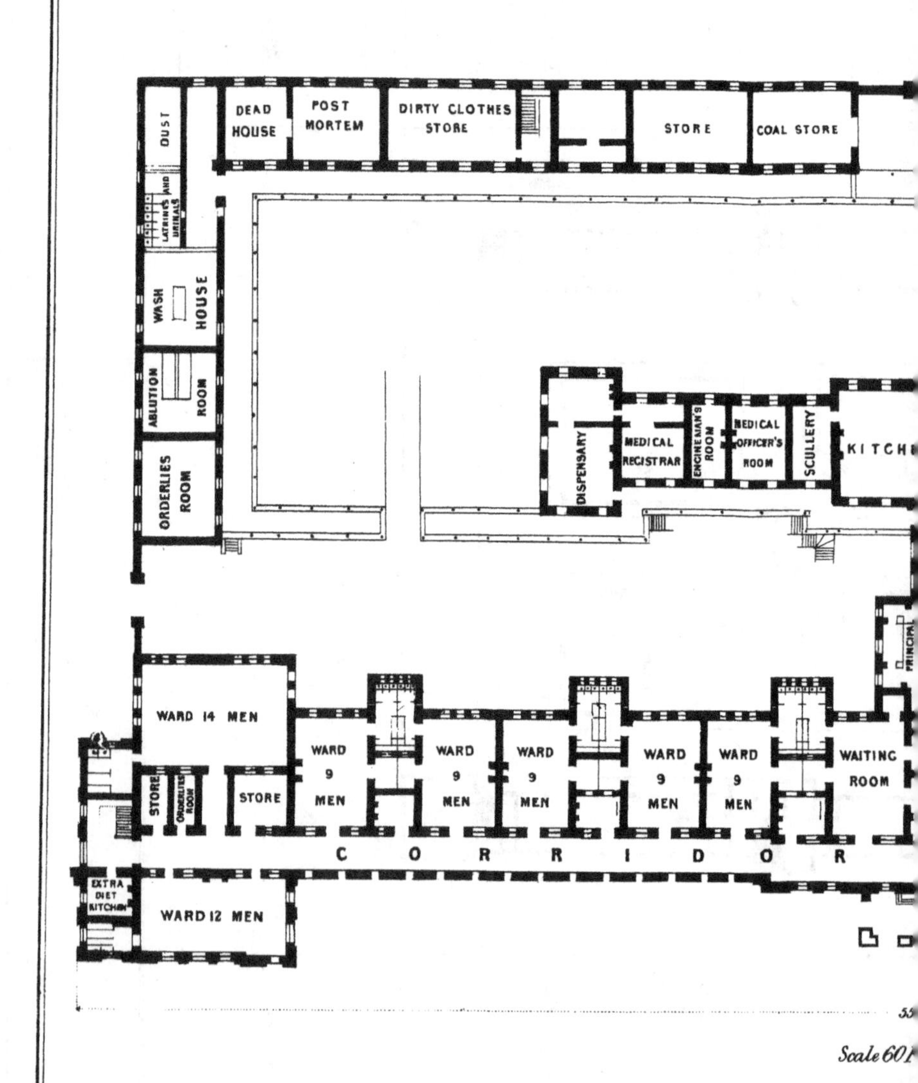

John W. Parker